Electronic Communication

AUTHORS

Dr. M. D. Selvaraj
Professor,
Department of ECE,
Indian Institute of Information Technology,
Design & Manufacturing, Kancheepuram,
Chennai, Tamil Nadu

Dr. Prabagarane Nagaradjane
Associate Professor,
Department of ECE,
Sri Sivasubramania Nadar College of Engineering,
Kalavakkam, Chennai

REVIEWER

Dr. Rishi Raj Sharma
Assistant Professor
Defence Institute of Advanced Technology (DIAT),
Pune

KHANNA BOOK PUBLISHING CO. (P) Ltd.
4C/4344, Ansari Road, Darya Ganj,
New Delhi, India

All India Council for Technical Education
Nelson Mandela Marg, Vasant Kunj,
New Delhi, India

Dear Readers,

To prevent the piracy, this book is secured with HIGH SECURITY HOLOGRAM on the front title cover. In case you don't find the hologram on the front cover title, please write us to at contact@khannabooks.com or whatsapp us at +91-99109 09320 and avail special gift voucher for yourself.

Specimen of Hologram on front Cover title:

Moreover, there is a SPECIAL DISCOUNT COUPON for you with EVERY HOLOGRAM.

How to avail this SPECIAL DISCOUNT:

Step 1: Scratch the hologram

Step 2: Under the scratch area, your "coupon code" is available

Step 3: Logon to www.khannabooks.com

Step 4: Use your "coupon code" in the shopping cart and get your copy at a special discount

Step 5: Enjoy your reading!

ISBN: 978-93-5538-269-6
Book Code: DIP457EN

Electronic Communication

by Dr. M. D. Selvaraj and
Dr. Prabagarane Nagaradjane

First Edition: 2025

Originally Published by:
All India Council for Technical Education, Ministry of Education, GOI

Exclusively Published & Distributed by:
Khanna Book Publishing Co. (P) Ltd.
4C/4344, Ansari Road, Daryaganj,
New Delhi (INDIA)

Visit us at: **www.khannabooks.com**
Write us at: **contact@khannabooks.com**

To view complete list of books, Please scan the QR Code:	KPH

Printed in Bharat.

प्रो. टी. जी. सीताराम
अध्यक्ष
Prof. T. G. Sitharam
Chairman

सत्यमेव जयते

आज़ादी का अमृत महोत्सव

अखिल भारतीय तकनीकी शिक्षा परिषद्
(भारत सरकार का एक सांविधिक निकाय)
(शिक्षा मंत्रालय, भारत सरकार)
नेल्सन मंडेला मार्ग, वसंत कुंज, नई दिल्ली–110070
दूरभाष : 011–26131498
ई–मेल : chairman@aicte-india.org

ALL INDIA COUNCIL FOR TECHNICAL EDUCATION
(A STATUTORY BODY OF THE GOVT. OF INDIA)
(Ministry of Education, Govt. of India)
Nelson Mandela Marg, Vasant Kunj, New Delhi-110070
Phone : 011-26131498
E-mail : chairman@aicte-india.org

FOREWORD

Engineers are the backbone of any modern society. They are the ones responsible for the marvels as well as the improved quality of life across the world. Engineers have driven humanity towards greater heights in a more evolved and unprecedented manner.

The All India Council for Technical Education (AICTE), have spared no efforts towards the strengthening of the technical education in the country. AICTE is always committed towards promoting quality Technical Education to make India a modern developed nation emphasizing on the overall welfare of mankind.

An array of initiatives has been taken by AICTE in last decade which have been accelerated now by the National Education Policy (NEP) 2020. The implementation of NEP under the visionary leadership of Hon'ble Prime Minister of India envisages the provision for education in regional languages to all, thereby ensuring that every graduate becomes competent enough and is in a position to contribute towards the national growth and development through innovation & entrepreneurship.

One of the spheres where AICTE had been relentlessly working since past couple of years is providing high quality original technical contents at Under Graduate & Diploma level prepared and translated by eminent educators in various Indian languages to its aspirants. For students pursuing 2nd year of their Engineering education, AICTE has identified 87 books, which shall be translated into 12 Indian languages - Hindi, Tamil, Gujarati, Odia, Bengali, Kannada, Urdu, Punjabi, Telugu, Marathi, Assamese & Malayalam. In addition to the English medium, books in different Indian Languages are going to support the students to understand the concepts in their respective mother tongue.

On behalf of AICTE, I express sincere gratitude to all distinguished authors, reviewers and translators from the renowned institutions of high repute for their admirable contribution in a record span of time.

AICTE is confident that these outcomes based original contents shall help aspirants to master the subject with comprehension and greater ease.

(Prof. T. G. Sitharam)

ACKNOWLEDGEMENT

The authors are grateful to the authorities of AICTE, particularly Prof. T. G. Sitharam, Chairman; Prof. Abhay Jere, Vice-Chairman; Prof. Rajive Kumar, Member-Secretary and Dr. Sunil Luthra, Director, Training and Learning Bureau, for their meticulous planning to publish the book on **Electronic Communication**. We sincerely acknowledge the valuable contributions of the book reviewer Dr. Rishi Raj Sharma, Assistant Professor, Defence Institute of Advanced Technology (DIAT), Pune, for making it student-friendly and artistically giving a better shape.

The first author, M. D. Selvaraj expresses his deepest gratitude to Prof. M. V. Karthikeyan, Director, IIITDM Kancheepuram. His leadership, guidance, support, and encouragement have empowered him to collaborate effectively and achieve the goals. M. D. Selvaraj would like to express his deepest gratitude to his wife M.S. Jamuna, and children Kaarthikeyan and Sahana for their unwavering love, support, and understanding throughout the writing of this book. Their encouragement and belief have been the greatest sources of strength.

The second author, Prabagarane Nagaradjane would like to express his deep sense of gratitude to the management of SSN. Furthermore, Prabagarane Nagaradjane would like to express his earnest appreciation to his parents Nagaradjane V and Sarasvady N and his wife Gayathri M, and his kids Sayandicaa and Vibileiche for their everlasting love, assistance, and comfort in completing this book project. Just a few words mentioned here can never fully capture all his appreciation for his family members for the encouragement they have given him right through during the entire course of writing this book.

Also, special thanks to Jaganathan Rajapandi for supporting the authors with his excellent typesetting expertise.

This book is the outcome of various suggestions from AICTE members, experts, and authors who shared their opinions and thoughts on developing engineering education further in our country. Acknowledgments are due to the contributors and different workers in this field whose published books, review articles, papers, photographs, footnotes, references, and other valuable information enriched us while writing the book.

M. D. Selvaraj
Prabagarane Nagaradjane

PREFACE

The book Electronic Communication results from the rich experience of teaching introductory communication courses. This book aims to expose the basic concepts of communication engineering and the fundamentals of electronic communication to diploma students and enable them to get an insight into the subject. To keep in mind the purpose of comprehensive coverage and to provide essential supplementary information, we have included the topics recommended by the AICTE in a very systematic and orderly manner throughout the book. We tried to present the subject's fundamental concepts in the simplest possible way.

While preparing the manuscript, we considered the standard textbooks and developed sections like exercise questions and problems. While preparing the different units, we emphasize presenting definitions and a comprehensive synopsis of formulae for a quick revision of the basic principles. The book covers all types of primary and advanced-level problems, and these have been presented in a very logical and systematic manner.

We have included the relevant laboratory practical experiments throughout the book to ensure our readers get hands-on training on the topics discussed. It is a salient feature of this book. In addition, besides some essential information for the users under the heading 'Know More', we have clarified some crucial basic information in the annexure section.

Regarding this book, we planned to provide a thorough grounding in analog and pulse communications on the topics covered. The subject matters are presented constructively to prepare diploma students to work in different electronic industries at the forefront of technology.

We hope that this book will inspire the students to learn and discuss the ideas behind the basic principles of electronic communication and will undoubtedly contribute to developing a solid foundation for the subject. We appreciate all beneficial comments and suggestions that will contribute to the improvement of the future editions of the book. It gives us immense pleasure to place this book in the hands of the teachers and students. Working on different aspects covered in the book was a great pleasure.

M. D. Selvaraj
Prabagarane Nagaradjane

OUTCOME BASED EDUCATION

For the implementation of an outcome-based education, the first requirement is to develop an outcome-based curriculum and incorporate an outcome based assessment in the education system. By going through outcome-based assessments, evaluators will be able to evaluate whether the students have achieved the outlined standard, specific and measurable outcomes. With the proper incorporation of outcome-based education, there will be a definite commitment to achieve a minimum standard for all learners without giving up at any level. At the end of the programme running with the aid of outcome-based education, a student will be able to arrive at the following outcomes:

Programme Outcomes (POs) are statements that describe what students are expected to know and be able to do upon graduating from the programme. These relate to the skills, knowledge, analytical ability attitude and behaviour that students acquire through the programme. The POs essentially indicate what the students can do from subject-wise knowledge acquired by them during the programme. As such, POs define the professional profile of an engineering diploma graduate.

The National Board of Accreditation (NBA) has defined the following seven POs for an Engineering diploma graduate:

PO1. Basic and discipline-specific knowledge: Apply knowledge of basic mathematics, science and engineering fundamentals and engineering specialization to solve the engineering problems.

PO2. Problem analysis: Identify and analyses well-defined engineering problems using codified standard methods.

PO3. Design/development of solutions: Design solutions for well-defined technical problems and assist with the design of systems components or processes to meet specified needs.

PO4. Engineering tools, experimentation and testing: Apply modern engineering tools and appropriate technique to conduct standard tests and measurements.

PO5. Engineering practices for the society, sustainability and environment: Apply appropriate technology in context of society, sustainability, environment and ethical practices.

PO6. Project management: Use engineering management principles individually, as a team member or a leader to manage projects and effectively communicate about well-defined engineering activities.

PO7. Life-long learning: The ability to analyse individual needs and engage in updating in the context of technological changes.

COURSE OUTCOMES (CO)

By the end of the course, the students will be able to:

CO-1: Use different modulation and demodulation techniques used in analog communication.

CO-2: Identify and solve basic communication problems.

CO-3: Analyse transmitter and receiver circuits.

CO-4: Compare and contrast the design issues, advantages, disadvantages and limitations of analog communication systems.

CO-5: Understand the working of spread spectrum communication system and analyse its performance

CO-6: Able to demonstrate the practical applications through experiments.

Mapping of course outcomes with POs to be done according to the matrix given below:

Course Outcomes	EXPECTED MAPPING WITH PROGRAMME OUTCOMES (1- Weak Correlation; 2- Medium correlation; 3- Strong Correlation)						
	PO-1	PO-2	PO-3	PO-4	PO-5	PO-6	PO-7
CO-1	3	3	3	1	1	2	3
CO-2	3	3	3	2	1	2	3
CO-3	3	2	3	3	1	2	3
CO-4	3	2	2	2	1	2	3
CO-5	3	2	2	2	1	2	3
CO-6	3	2	2	2	2	3	3

GUIDELINES FOR TEACHERS

To implement Outcome Based Education (OBE) knowledge level and skill set of the students should be enhanced. Teachers should take a major responsibility for the proper implementation of OBE. Some of the responsibilities (not limited to) for the teachers in OBE system may be as follows:

- Within reasonable constraint, they should manoeuvre time to the best advantage of all students.
- They should assess the students only upon certain defined criterion without considering any other potential ineligibility to discriminate them.
- They should try to grow the learning abilities of the students to a certain level before they leave the institute.
- They should try to ensure that all the students are equipped with the quality knowledge as well as competence after they finish their education.
- They should always encourage the students to develop their ultimate performance capabilities.
- They should facilitate and encourage group work and team work to consolidate newer approach.
- They should follow Blooms taxonomy in every part of the assessment.

Bloom's Taxonomy

Level	Teacher should Check	Student should be able to	Possible Mode of Assessment
Creating	Students ability to create	Design or Create	Mini project
Evaluating	Students ability to Justify	Argue or Defend	Assignment
Analysing	Students ability to distinguish	Differentiate or Distinguish	Project/Lab Methodology
Applying	Students ability to use information	Operate or Demonstrate	Technical Presentation/ Demonstration
Understanding	Students ability to explain the ideas	Explain or Classify	Presentation/Seminar
Remembering	Students ability to recall (or remember)	Define or Recall	Quiz

GUIDELINES FOR STUDENTS

Students should take equal responsibility for implementing the Outcome Based Education (OBE). Some of the responsibilities (not limited to) for the students in OBE system are as follows:

- Students should be well aware of each Unit Outcome (UO) before the start of a unit in each and every course.
- Students should be well aware of each Course Outcome (CO) before the start of the course.
- Students should be well aware of each Program Outcome (PO) before the start of the program.
- Students should think critically and reasonably with proper reflection and action.
- Learning of the students should be connected and integrated with practical and real life consequences.
- Students should be well aware of their competency at every level of OBE.

ABBREVIATIONS AND SYMBOLS

List of Abbreviations

General Terms			
Abbreviations	**Full form**	**Abbreviations**	**Full form**
AM	Amplitude modulation	AF	Amplify and forward
DSB-SC	Double sideband – suppressed carrier	RFC	Radio-frequency choke
SSC-SC	Single sideband – supressed carrier	PLL	Phase-locked loop
VSB	Vestigial side band	VCO	Voltage-controlled oscillator
FM	Frequency modulation	ADC	Analog-to-digital converter
NBFM	Narrow band frequency modulation	SNR	Signal-to-noise ratio
WBFM	Wide-band frequency modulation	DAC	Digital-to-analog converter
AWGN	Additive white Gaussian noise	SQR	Signal-to-quantized noise ratio
PAM	Pulse amplitude modulation	UHF	Ultra-high frequency
PWM	Pulse-width modulation	FET	Field effect transistor
PPM	Pulse position modulation	NRZ	Non-return to zero
PCM	Pulse code modulation	RZ	Return to zero
DPCM	Differential pulse code modulation	DC	Direct current
		AMI	Alternate mark inversion
DM	Delta modulation	ISI	Inter-symbol interference
ADM	Adaptive delta modulation	PN	Pseudo-noise
ASK	Amplitude shift keying	SS	Spread spectrum
FSK	Frequency shift keying	LAN	Local area network
PSK	Phase shift keying	CDMA	Code division multiple access
MSK	Minimum shift keying	DSSS	Direct sequence spread spectrum
QAM	Quadrature amplitude modulation	FHSS	Frequency hop spread spectrum
BPSK	Binary phase shift keying	PG	Processing gain
QPSK	Quadrature phase shift keying	FH/MFSK	Frequency hopping/M-ary frequency shift keying
8-PSK	8-phase shift keying		
TRF	Tuned radio-frequency	OSS	Orthogonal spread spectrum
RF	Radio-frequency	LPI	Low probability of intercept
IF	Intermediate frequency	SSMA	Spread-spectrum multiple access
EM	Electromagnetic		

List of Units and Symbols

Symbols	Description	Symbols	Description
A_m	Amplitude of the modulating wave	β	Deviation factor of the FM wave
f_m	Frequency of the modulating wave	B	Carson's bandwidth of FM wave
A_c	Amplitude of the carrier wave	C_n	Fourier series constant term
f_c	Frequency of the carrier wave	J_n	Bessel function
C	Velocity of light	n	Total number of potential values
f	Frequency of a wave	f_s	Sampling frequency
λ	Wavelength of a Wave	T_s	Sampling time period
K_a	Sensitivity factor of the AM wave	Z	Impedance value
μ	Modulation index of the AM wave	C	Capacitance value
A_{max}	Maximum amplitude of the modulated wave	Q	Quantization error
A_{min}	Minimum amplitude of the modulated wave	q	Sample value of Q
B_{AM}	Bandwidth of the AM wave	Δ	Step size
P_c	Power of the carrier wave	L	Number of representation levels
P_{LSB}	Power of the lower side band	R	Number of bits per sample
P_{USB}	Power of the upper side band	m_{max}	Maximum amplitude of the analog signal
A_T	Total power consumption	Qe	Quantization error
A_{car}	Amplitude of the carrier wave	ti	Sampling time
A_{LSB}	Amplitude of the lower side band	Tb	Time period of a bit
A_{USB}	Amplitude of the upper side band	Rb	Bit rate
R_L	Load resistance	A	Roll-off factor
τ	Time delay	N	Period of maximum length sequence
f_{LO}	Frequency of the local oscillator	Rc	Chip rate
f_{RF}	Frequency of the radio-frequency wave	M	Number of symbols
f_{IF}	Intermediate frequency of the wave	Rh	Hop rate
K_f	Sensitivity factor of the FM wave		
Δ_f	Frequency deviation of the FM wave		

LIST OF FIGURES

CONTENTS

1 Amplitude and Frequency Modulation

UNIT SPECIFICS

Through this unit, we will discuss the following aspects:

- Elements of a communication system
- Need for modulation
- DSB-FC: the modulation index, spectral analysis and modulator and demodulator circuits
- DSB-SC: modulator and demodulator
- Envelope detector and coherent demodulator techniques
- SSB-SC: frequency and phase discrimination methods
- VSB
- Tuned radio-frequency and Superheterodyne receiver
- FM
- Compare the AM and FM and also study the applications of them
- Study various AM transmitters and receivers

The practical applications of the various modulation techniques are discussed for generating further curiosity and creativity as well as improving problem-solving capacity.

Besides giving questions of short and long answer types marked in two categories following lower and higher order of Bloom's taxonomy, assignments through a number of numerical problems, a list of references and suggested readings are given in the unit so that one can go through them for practice. It is important to note that for getting more information on various topics of interest some websites are provided at the end, which can be clicked for relevant supportive knowledge.

After the related practical, there is a content-based 'Know More' section. This section has been carefully designed in such a way that the supplementary information provided in this part becomes beneficial for the users of the book. This section mainly highlights the initial activity, examples of some interesting facts, analogy, history of the development of the subject focusing the salient observations and finding, timelines starting from the development of the concerned topics up to recent times, applications of the subject matter for our day-to-day real life or/and industrial applications on variety of aspects, case study related to environmental, sustainability, social and ethical issues whichever applicable, and finally inquisitiveness and curiosity topics of the unit.

RATIONALE

This unit on amplitude and frequency modulation begins with the introduction of a communication system. The modulator is an important block in the transmitter. Therefore, the unit answers the questions of what is modulation and need for modulation. The overview about various techniques is provided by the classification chart with applications. The spectrum and power analysis of DSB-FC, DSB-SC are provided in depth to understand the subject. The envelope detector and coherent demodulator circuits are explained in detail. The spectral efficient and power efficient SSB-SC technique is generated by frequency and phase discriminator methods.

Low-power and high-power AM transmitters are discussed at length with necessary block diagrams. Followed by tuned radio-frequency and superheterodyne receiver are explained in depth. The spectrum analysis of FM is illustrated at length to find the bandwidth of the FM signal. The generation and detection of FM is discussed in detail with necessary block diagram.

PRE-REQUISITES

Mathematics and Physics: Class IX and Class X.

UNIT OUTCOMES

List of outcomes of this unit is as follows:

U1-O1: Definition of modulation and need for modulation

U1-O2: Generation and detection of various modulation techniques such as DSB-FC, DSB-SC, SSB-SC, VSB, and FM

U1-O3: Analyse the modulation index, bandwidth and power of various modulation techniques

U1-O4: Study various AM transmitters and receivers

U1-O5: Compare the AM and FM and study the applications of them

Unit-1 Outcome	EXPECTED MAPPING WITH COURSE OUTCOMES (1 - Weak Correlation; 2 - Medium Correlation; 3 - Strong Correlation)					
	CO-1	CO-2	CO-3	CO-4	CO-5	CO-6
U1-O1	3	-	-	-	-	1
U1-O2	1	3	-	-	-	2
U1-O3	1	1	3	-	-	2
U1-O4	1	3	-	-	-	1
U1-O5	1	1	1	-	-	1

Communication System

1.1 INTRODUCTION

Electronic communication is the process of transferring information from a source terminal to a destination. The communication system as shown in Fig. 1.1 is essentially consisting of five blocks.

(i) The message is stored in the information source. The message may be in the form of text, speech, image, or video.

(ii) The transmitter receives the message signal from the information source and transforms the message in such way that it travels in the channel safely. The transmitter has several blocks such as modulator, channel coding, encryption etc.

Fig. 1.1: Block diagram of a communication system

(iii) The transmission medium is referred as channel. The transmission is generally divided into wired and wireless channel. The example for wired channel is copper wire, coaxial cable, and optical fibre. The example for wireless channel is Free space (air).

(iv) The receiver is counterpart of the transmitter block. The original message is recovered from the noisy received signal. The receiver has demodulator, channel decoder, decryption, etc.

(v) The destination is the intended terminal or user to receive the original message.

1.2 MODULATION

Basically, modulation is a frequency translation process in which a low-frequency message signal is shifted to the higher frequency carrier signal range. A few features (amplitude or frequency) of the higher frequency signal are modified based on the message signal. The process of modulation and demodulation is shown in Fig. 1.2.

There are three signals in the modulator section:

(i) Message signal: It is usually a low-frequency signal carries useful information. The message signal may be represented as

$$m(t) = A_m \cos(2\pi f_m t) \tag{1.1}$$

where A_m is the amplitude of the message signal and f_m is the frequency of the message signal.

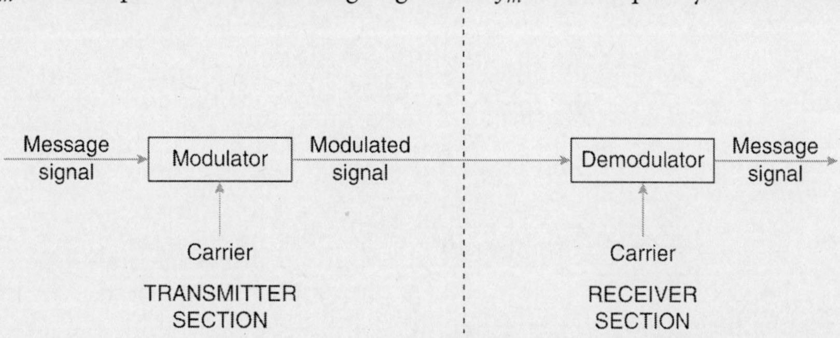

Fig. 1.2: Process of modulator and demodulator in a communication system

(ii) Carrier signal: It is a high-frequency signal carries the message signal to the longer distance. The carrier signal is represented as,

$$c(t) = A_c \cos(2\pi f_c t) \tag{1.2}$$

where A_c is the amplitude of the carrier signal and f_c is frequency of the carrier signal.

Usually, the carrier frequency chosen is always greater than the message signal frequency, i.e. $f_c >> f_m$.

(iii) Modulated signal is the output of the modulator, which is transmitted to the destination through the channel.

1.3 NEED FOR MODULATION

The modulation process is extremely important in the communication system due to the following reasons.

1.3.1 To Reduce the Antenna Height

For efficient transmission, the transmitting antenna should have length at least equal to a quarter of the wavelength of the signal to be transmitted. The effect of modulation can be observed is two cases as follows:

Case (i): *Without Modulation*:

Suppose a message signal is transmitted without modulation, then voice signal = 3 kHz {K = 10^3}. We know that,

$$\text{Wavelength } \lambda = \frac{\textit{velocity of light}(c)}{\textit{frequency}(f)}$$

$$= \frac{3 \times 10^8 \text{ m/s}}{3 \times 10^3} = 10^5 \text{ m} = 100 \text{ km}$$

$$\text{Vertical antenna height} \approx \frac{\lambda}{4} = \frac{100 \text{ km}}{4} = 25 \text{ km} \tag{1.3-a}$$

Case (ii): *With Modulation*:

Suppose, the message signal is modulated by a carrier signal having frequency $f_c = 300 \text{ MHz} (M = 10^6)$

then

$$\text{Wavelength } \lambda = \frac{\textit{velocity of light}(c)}{\textit{frequency}(f)}$$

$$\lambda = \frac{3 \times 10^8 \text{ m/s}}{300 \times 10^6} = 1 \text{ m}$$

$$\text{Vertical antenna height} \approx \frac{\lambda}{4} = \frac{1 \text{ m}}{4} = 0.25 \text{ m} \tag{1.3-b}$$

From Eq. (1.3-a) and (1.3-b), it is understood that it is difficult to construct a practical antenna without modulation.

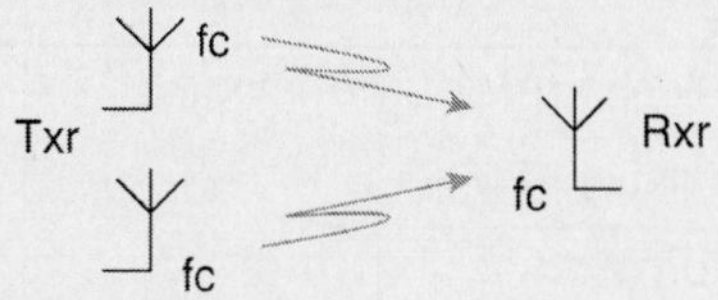

Fig. 1.3: Two-antenna example

1.3.2 To Avoid Interference

Interference refers to the process of disruption in which the intended signal is modified by an unwanted signal.

If two antennas of different communication system use the same carrier frequency or closely related carrier frequency then they tend to interfere with each other (Fig. 1.3). Therefore, to overcome this problem, transmitted signal should be modulated by different carrier signals.

1.3.3 To Avoid Poor Radiation

The power radiated from the antenna is a direct function of the length of the antenna and inverse function to the wavelength of the carrier wave.

The power radiated by an antenna can be increased by decreasing the wavelength of the carrier which is achieved by increasing its carrier frequency. This is performed through a modulation process.

1.3.4 To Effectively Utilize the Transmission Medium

Without modulation, if more than one signal is transmitted through a cable, then those signals mix-up among them. In such a case, the received signal can't be recovered separately. To overcome this difficulty, modulation may be used to separate message signals into different spectral bands.

Modulation techniques are categorized into three classes, namely the analog, digital, and pulse modulation. A complete tree diagram of various modulation techniques with their branches is given in Fig. 1.4.

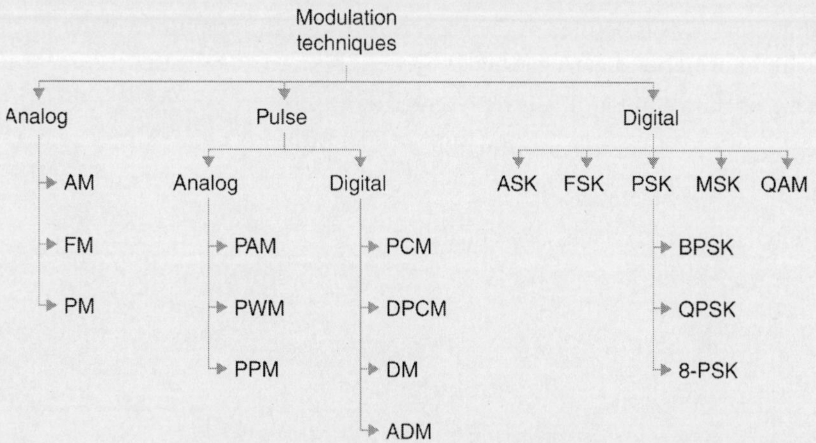

Fig. 1.4: Various modulation techniques

Modulation Techniques		Message Signal	Carrier Signal	Modulated Signal
Analog	Amplitude modulation (AM)	Analog	Analog	Analog
	Frequency modulation (FM)			
	Phase modulation (PM)			
Pulse-Analog	Pulse Amplitude modulation (PAM)	Analog	Pulse	Analog
	Pulse width modulation (PWM)			
	Pulse position modulation (PPM)			
Pulse-Digital	Pulse code modulation (PCM)	Analog	Pulse	Digital
	Differential pulse code modulation (DPCM)			
	Delta modulation (DM)			
	Adaptive delta modulation (ADM)			
Digital	Amplitude shift keying (ASK)	Digital	Analog	Analog
	Frequency shift keying (FSK)			
	Phase shift keying (PSK)			
	Minimum shift keying (MSK)			
	Quadrature amplitude modulation (QAM)			

1.4 AMPLITUDE MODULATION

Amplitude modulation (AM) is defined as the process by which the amplitude of the carrier is varied in accordance with an information-bearing signal.

The AM wave, as shown in Fig. 1.5, is represented as

$$x_{AM}(t) = \left(1 + K_a m(t)\right) A_c \cos\left(2\pi f_c t\right) \quad (1.4)$$

where

K_a – sensitivity factor, $m\ (t)$ – message signal.

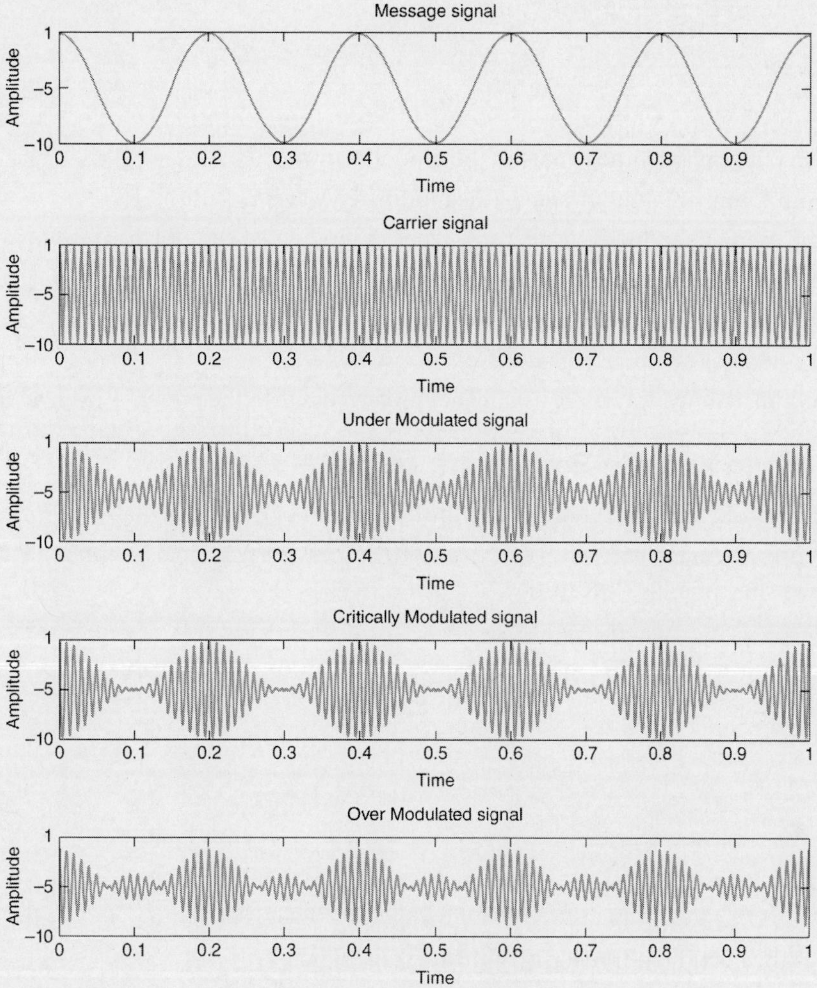

Fig. 1.5: AM waveforms

1.4.1 Modulation Index

The modulation index of the AM is the ratio of the maximum amplitude of the message signal to the maximum amplitude of the carrier.

$$\mu = \frac{A_m}{A_c} \tag{1.5}$$

Alternatively, the modulation index is also defined as

$$\mu = K_a \left|\min\left(m(t)\right)\right| \tag{1.6}$$

where $|\cdot|$ represents mod operation which always gives positive value. If $m(t) = A_m \cos(2\pi f_m t)$ then $\mu = K_a\left|(-A_m)\right| = K_a A_m.(\because \cos(0) = 1)$.

The modulation index is also obtained from the AM waveform,

$$\mu = \frac{A_{max} - A_{min}}{A_{max} + A_{min}} \tag{1.7}$$

where

A_{max} – the maximum amplitude value of the modulated wave

A_{min} – the minimum amplitude value of the modulated wave

Based on the modulation index value, three types of modulated waveforms are generated.

when μ < 1, then it is called *under modulation.*

when μ < 1, then it is called *critical modulation.*

when μ < 1, then it is called *over modulation.*

The over modulated AM waves are usually not preferred due to the presence of envelope distortion.

1.4.2 Spectrum and the Bandwidth of AM

The spectral analysis of AM is useful in finding the bandwidth of the AM signal.

Consider a message signal $m(t)$ with spectrum $M(f)$. The message signal is represented in frequency domain as follows shown in Fig. 1.6.:

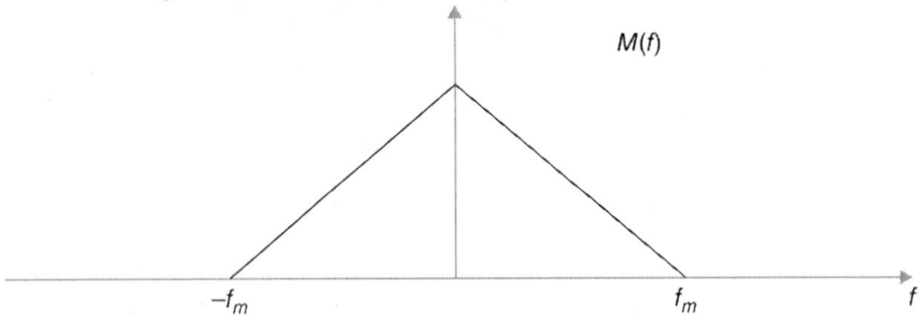

Fig. 1.6: Spectrum of the message signal

where f_m is the maximum frequency of the message signal.

The AM wave is represented as

$$x_{AM}(t) = A_c\left(1 + K_a m(t)\right)\cos(2\pi f_c t)$$

$$= A_c \cos(2\pi f_c t) + A_c K_a m(t)\cos(2\pi f_c t). \tag{1.8}$$

By Euler's formula, we can write

$$\cos(2\pi f_c t) = \left[\frac{e^{j2\pi f_c t} + e^{-j2\pi f_c t}}{2}\right]. \tag{1.9}$$

The Fourier transform $e^{\pm j2\pi f_c t}$ of is $\delta(f \mp f_c)$, where $\delta(f)$ is the impulse function defined as $\delta(f) = 1$ when $f = 0$.

Therefore, we can write Fourier transform pair as

$$A_c \cos(2\pi f_c t) \leftrightarrow \frac{A_c}{2}\left[\delta(f - f_c) + \delta(f + f_c)\right]. \tag{1.10}$$

Now, let us focus on the second term of Eq. (1.8), i.e. $A_cK_am(t)\cos(2\pi f_ct)$., We know that from Fourier transform property, multiplication in time-domain gives convolution in a frequency domain. The Fourier transform $A_cK_am(t)$ of is $A_cK_aM(f)$. Therefore, we represent $A_cK_am(t)\cos(2\pi f_ct)$ part in the frequency domain as,

$$= A_cK_aM(f) * \frac{1}{2}[\delta(f-f_c)+\delta(f+f_c)] \tag{1.11}$$

where * represents the convolution operator.

$$= \frac{A_cK_a}{2}[M(f)*\delta(f-f_c)+M(f)*\delta(f+f_c)]. \tag{1.12}$$

It can be simplified as,

$$= \frac{A_cK_a}{2}[M(f-f_c)+M(f+f_c)] \tag{1.13}$$

By combining Eq. (1.10) and (1.13) and substituting in Eq. (1.8) results in,

$$X_{AM}(f) = \frac{A_c}{2}[\delta(f-f_c)+\delta(f+f_c)] + \frac{A_cK_a}{2}[M(f-f_c)+M(f+f_c)] \tag{1.14}$$

Bandwidth is the band of frequencies and usually difference between the highest frequency component and lowest frequency component.

Bandwidth, B = $fH - f_L$

Bandwidth of the AM signal, $B_{AM} = (f_c + f_m) - (f_c - f_m)$

$$B_{AM} = 2f_m. \tag{1.15}$$

The bandwidth of the AM is twice the frequency of the message signal.

1.4.3 Power of the AM Signal

The total power in the AM signal is the sum of the carrier power (P_c) component and power in the upper sideband (P_{USB}) and lower sideband (P_{LSB}) components.

Consider a message signal

$$m(t) = A_m\cos(2\pi f_mt). \tag{1.16}$$

The AM wave is

$$x_{AM}(t) = A_C\left(1 + k_aA_M\cos(2\pi f_mt)\right)\cos(2\pi f_ct). \tag{1.17}$$

Substituting modulation index, $\mu = k_aA_M$ in Eq. (1.17) results in the expression $x_{AM}(t)$, which is written as,

$$x_{AM}(t) = A_C\cos(2\pi f_ct) + \mu A_C\cos(2\pi f_mt)\cos(2\pi f_ct) \tag{1.18}$$

We know the trigonometric formula

$$\cos(f_1)\cos(f_2) = \frac{\cos(f_1+f_2)+\cos(f_1-f_2)}{2}. \tag{1.19}$$

Applying Eq. (1.19) in (1.18) results in

$$= \underbrace{A_C \cos(2\pi f_c t)}_{\text{Carrier}} + \underbrace{\frac{\mu A_C}{2}\cos(2\pi(f_c + f_m)t)}_{\text{LSB}} + \underbrace{\frac{\mu A_C}{2}\cos(2\pi(f_c - f_m)t)}_{\text{USB}} \quad (1.20)$$

The total power in the AM wave is

$$P_T = P_C + P_{LSB} + P_{USB}. \quad (1.21)$$

The power of a signal is ratio of the square of voltage/amplitude and resistor, $\left(\text{i.e., } P = \frac{A^2}{R}\right)$

$$P_T = \frac{A_{carrier}^2}{R} + \frac{A_{LSB}^2}{R} + \frac{A_{USB}^2}{R}. \quad (1.22)$$

Let us assume $R = 1$ for the sake of simplicity

$$P_T = A_{carrier}^2 + A_{LSB}^2 + A_{USB}^2. \quad (1.23)$$

The rms and peak voltage of the wave is related as $A_{rms} = \frac{A_{peak}}{\sqrt{2}}$. The equation can be rewritten as

$$\begin{aligned} P_T &= \left(\frac{A_C}{\sqrt{2}}\right)^2 + \left(\frac{A_{LSB}}{\sqrt{2}}\right)^2 + \left(\frac{A_{USB}}{\sqrt{2}}\right)^2 \\ &= \frac{A_C^2}{2} + \left(\frac{0.5\mu A_C}{\sqrt{2}}\right)^2 + \left(\frac{0.5\mu A_C}{\sqrt{2}}\right)^2 \\ &= \frac{A_C^2}{2} + \frac{1}{8}\mu^2 A_C^2 + \frac{1}{8}\mu^2 A_C^2 \\ P_T &= \frac{A_C^2}{2}\left(1 + \frac{\mu^2}{2}\right). \end{aligned} \quad (1.24)$$

From Eq. (1.18), the carrier power component, P_c is $\frac{A_c^2}{2}$. Therefore, total AM power becomes

$$P_T = P_c\left(1 + \frac{\mu^2}{2}\right). \quad (1.25)$$

In AM, 33% of the power lies in the sideband which carries the message signal, whereas 67% of the power is wasted in the carrier component. Thus, AM with full carrier is not an efficient technique.

1.4.4 AM Wave Generation

The circuit shown in Fig. 1.7 is a voltage-divider bias circuit. The circuit is supplied by a dc power, V_{CC}. The resistors R_1 and R_2 are voltage divider resistors and R_C and R_E are collector and emitter resistors, respectively. The capacitors C_{C1} and C_{C2} are coupling capacitors and C_b is by-pass capacitor. The modulating signal is fed at the emitter and carrier is fed at the base of the transistor. Without the message signal, the circuit acts as a linear class 'A' amplifier.

When the message signal at the emitter varies, the gain of the circuit modulates the carrier signal. The variations in the amplitude of the carrier with respect to the message signal produce the AM waveform. The AM output is taken across the load resistor, C_L.

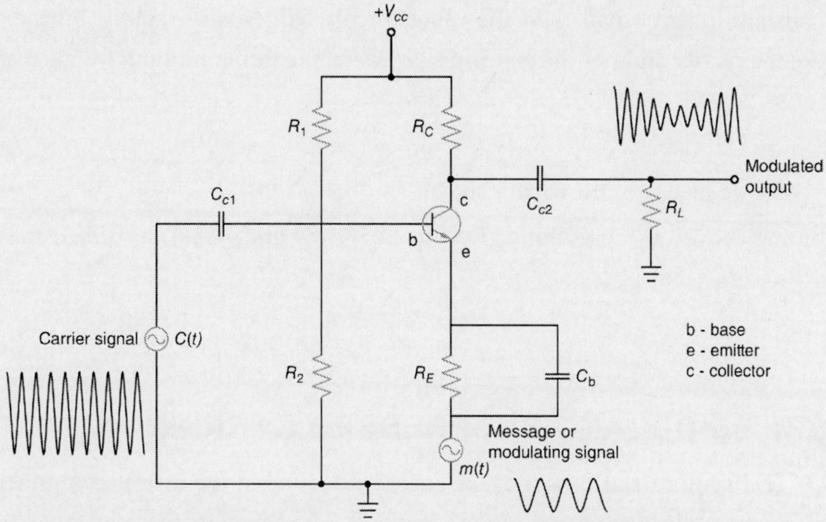

Fig. 1.7: Voltage- divider bias circuit for AM wave generation

1.4.5 Envelope Detector

The envelope detector traces the envelope of the AM modulated signal to recover the message signal. It works only when the modulated wave has no distortion. The circuit diagram of envelope detector is given in Fig. 1.8.

The AM wave is fed from a signal source. During positive half-cycle of the wave, the PN junction diode is forward biased, thus conducts the electric current. When $V_{in} > V_0$, the capacitor starts charging to the peak of the carrier. On the other hand, during negative half-cycle, when $V_{in} < V_0$, the diode is reverse-biased. The capacitor discharges slowly through resistor, R_L. The time constant τ, for the discharge of the capacitor is R_LC.

The choice of the τ is very important in tracking the envelope of the signal.

If time constant is very large then discharge is very slow then it fails to track the changes in amplitude. If message signal $m(t)$ varies faster then the amplitude of the modulated signal also varies faster. The frequency of the message signal is denoted by f_m thus we can write

$$\tau << \frac{1}{f_m} \tag{1.26}$$

That is, time period of the discharge of capacitor must be smaller to track variations.

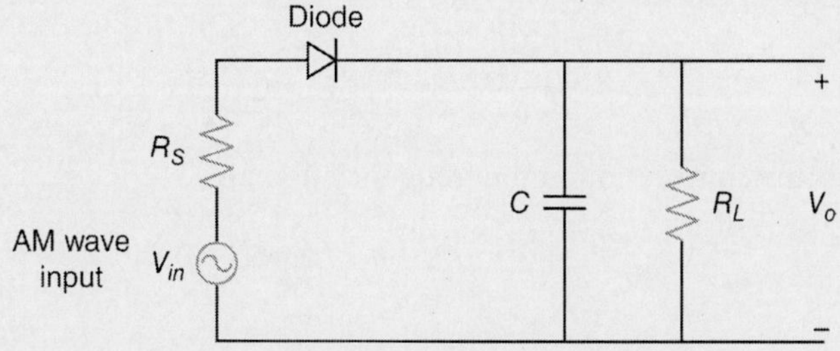

Fig. 1.8: Circuit diagram of an envelope detector

If time constant is very small then the capacitor discharges very rapidly. Further, output of the capacitor tracks the carrier and not the envelope. To overcome this condition, we must choose

$$\tau >> \frac{1}{f_c} \tag{1.27}$$

where f_c is the frequency of the carrier signal.

By combining above two conditions given in Eq. (1.26) and (1.27) on time constant τ, it can be written as,

$$\frac{1}{f_c} << \tau << \frac{1}{f_m} \tag{1.28}$$

1.5 DOUBLE SIDEBAND-SUPPRESSED CARRIER

In the case AM (DSB-full carrier), most of the energy is wasted in the carrier component. Further, the carrier (impulses at $\pm f_c$) does not carry any information. The complete message is carried by sideband. Therefore, the carrier power is suppressed to increase energy efficiency.

1.5.1 Generation of the DSB-SC Signal

The DSB-SC modulated signal is obtained by product of message signal and carrier signal as shown in Fig. 1.9, which is given below

$$x(t) = m(t)c(t) = m(t) \times A_c \cos(2\pi f_c t). \tag{1.29}$$

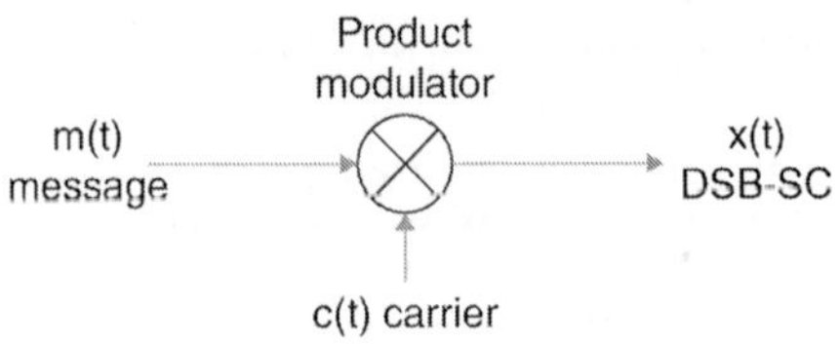

Fig. 1.9: Generation of the DSB-SC signal

1.5.2 Spectrum of the DSB-SC Signal

We know that the Fourier transform (FT) of $m(t)$ is $M(f)$. Also, FT of $A_c \cos(2\pi f_c t)$ is $\frac{A_c}{2}[\delta(f-f_c)+\delta(f+f_c)]$. Therefore, the spectrum of the DSB-SC is

$$X(f) = M(f) * \frac{A_c}{2}[\delta(f-f_c)+\delta(f+f_c)]$$

where * represents the convolution in the frequency domain.

$$= \frac{A_c}{2}[M(f)*\delta(f-f_c)+M(f)*\delta(f+f_c)]$$

$$X(f) = \frac{A_c}{2}[M(f-f_c)+M(f+f_c)] \tag{1.30}$$

which is shown in Fig. 1.10.

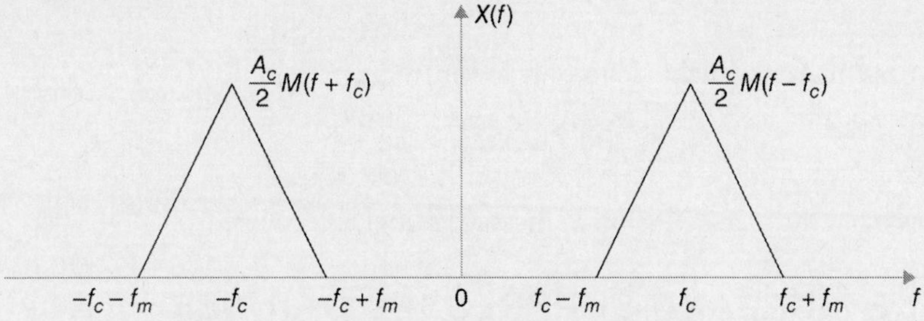

Fig. 1.10: Spectrum of the DSB- SC signal

1.5.3 Bandwidth of DSB-SC

$$B = (f_c + f_m) - (f_c - f_m)$$

$$B = 2f_m \tag{1.31}$$

The bandwidth of the DSB-SC is same as that of DSB-FC.

1.5.4 Coherent Demodulator

Another name of the coherent demodulator shown in Fig. 1.11 is the synchronous demodulator. It is because there is no phase difference between carrier wave of the incoming signal and locally generated carrier.

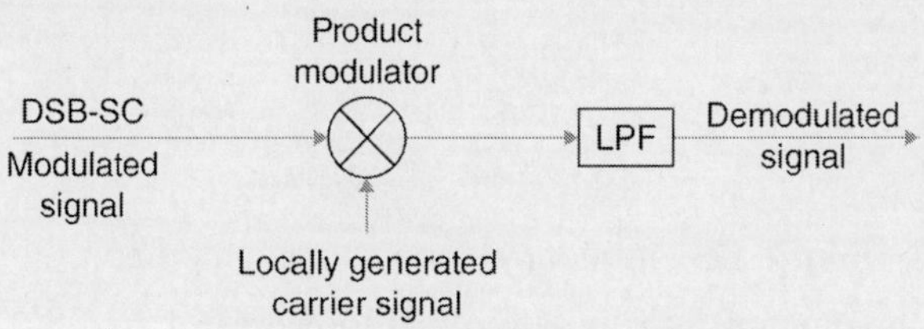

Fig. 1.11: Coherent demodulator

The incoming DSB-SC is represented as

$$x(t) = A_c m(t)\cos(2\pi f_c t).$$

The incoming DSB-SC signal is fed to the product modulator along with locally generated carrier $(\cos(2\pi f_c t))$. The output of the product modulator is

$$r(t) = A_c m(t)\cos(2\pi f_c t)\times\cos(2\pi f_c t) \tag{1.32}$$

$$= A_c m(t)\cos^2(2\pi f_c t)$$

$$= A_c m(t)\left[\frac{1+\cos(4\pi f_c t)}{2}\right]$$

$$r(t) = \frac{A_c m(t)}{2} + \frac{A_c}{2}m(t)\cos(4\pi f_c t). \tag{1.33}$$

The output of the product modulator is fed to the low-pass filter to remove higher frequencies ('$2f_c$' terms).

After low-pass filtering which allows only lower frequencies is,

$$r'(t) = \frac{A_c}{2} m(t) \tag{1.34}$$

which shows that the scaled version of message signal is recovered.

1.6 SINGLE SIDEBAND-SUPPRESSED CARRIER

In DSB-SC, both upper and lower sidebands carry the same information. That is, one of the sideband is redundant. Therefore, one sideband is removed and carrier power is suppressed in a technique called SSB-SC.

Generation of SSB-SC

There are two techniques used to generate the SSB-modulated signal.

(i) Frequency discrimination method

(ii) Phase discrimination method

1.6.1 Frequency Discrimination Method

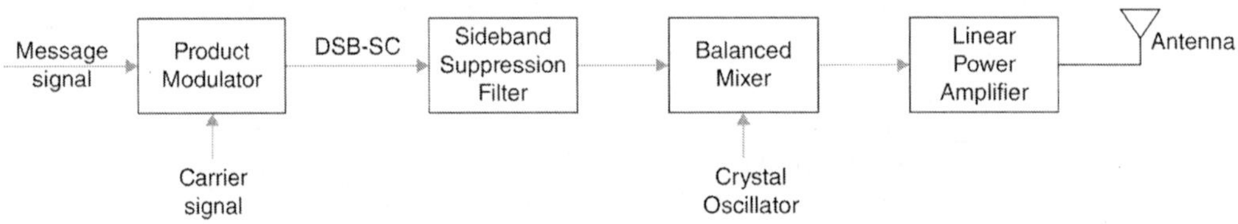

Fig. 1.12: Frequency discrimination method

The block diagram of Frequency discrimination method is shown in Fig. 1.12. There are two input signals to the product modulator: (i) message signal (ii) carrier signal which is generated from a crystal oscillator and produce stable oscillations. The product modulator produces DSB-SC modulated signal. Next, band-pass filters suppress the unwanted sideband and allows only one desired sideband. The sideband allowed by the filter may be the upper sideband (USB) or the lower sideband (LSB).

Then the SSB-SC signal is upconverted to the carrier frequency with the help of the balanced mixer and a crystal oscillator. The upconverted signal is amplified by a linear power amplifier and then is coupled to the antenna. The antenna converts the electrical signal to electromagnetic signal.

Advantages

(i) It is a simple circuit to produce SSB-SC.

(ii) Phase synchronization of carrier signal is not required.

Disadvantage

(i) It is difficult to obtain a sharp cut-off for the filters in practice.

1.6.2 Phase Discrimination Method

There are two identical product modulators (1 and 2) used in the generation of in-phase and quad-phase carrier signals as shown in Fig. 1.13. The wide-band modulating signal is fed directly to the product modulator-1.

The carrier signal, which has same frequency as that of the transmitted carrier signal, is generated from a crystal oscillator. The product modulator-1 receives both modulating signal and carrier signal to produces a DSB-SC which is given by

$$A_c A_m \cos(2\pi f_m t)\cos(2\pi f_c t). \tag{1.35}$$

Similarly, the product modulator-2 receives the 90° phase-shifted modulating signal from wide-band phase-shifter and 90° phase shifted carrier signal from crystal oscillator. The product modulator-2 produces DSB-SC, which is given by

$$A_c A_m \sin(2\pi f_m t)\sin(2\pi f_c t). \tag{1.36}$$

The summing amplifier adds both the DSB-SC signals given in Eq. (1.35) and (1.36) and produces SSB-SC output as follows:

$$x_{SSBSC}(t) = A_c A_m \cos(2\pi f_m t)\cos(2\pi f_c t) \pm A_c A_m \sin(2\pi f_m t)\sin(2\pi f_c t) \tag{1.37}$$

We know the trigonometric formula

$$\cos(A \mp B) = \cos A \cos B \pm \sin A \sin B \tag{1.38}$$

Apply this, we get

$$x_{SSBSC}(t) = A_c A_m \cos(2\pi(f_c \pm f_m)t) \tag{1.39}$$

where '–' indicates the LSB, and '+' indicates the USB.

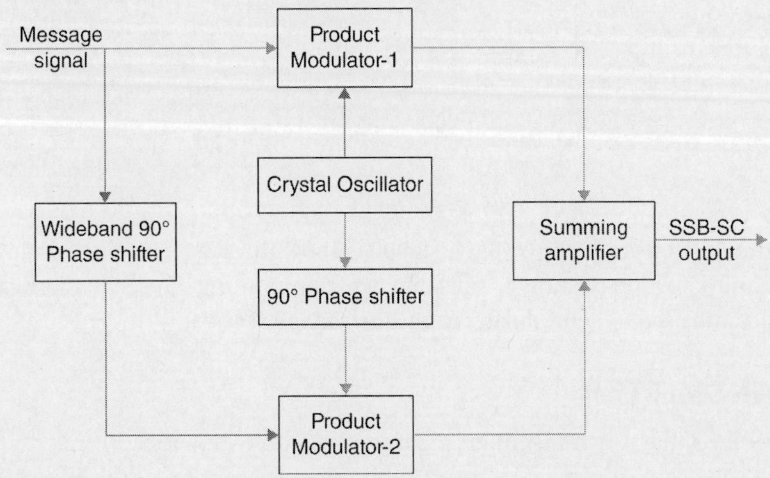

Fig. 1.13: Phase discrimination method

The output SSB-SC will have either LSB or USB depends upon the sign of the summing amplifier.

Advantages

(i) A sharp cut-off filter is not required.

(ii) It produces only one sideband without residue of other sideband.

Disadvantages

(i) The phase synchronization is required at the product modulators with the incoming carrier signal.

1.7 VESTIGIAL SIDEBAND MODULATION

In VSB, a band of frequencies, which consists of one sideband and a portion of another sideband (tail or vestige), is transmitted. The spectrum of TV transmission is shown in Fig. 1.14.

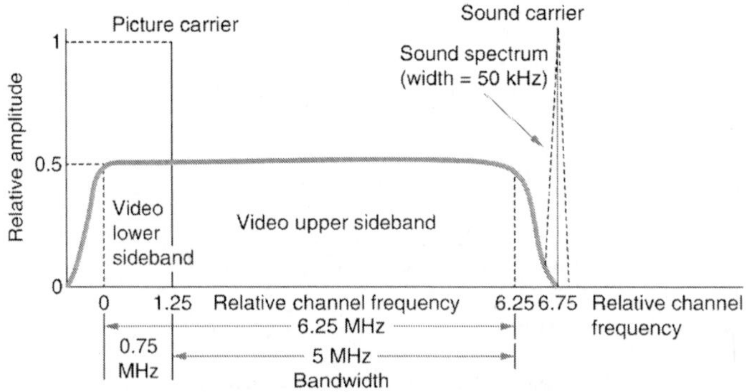

Fig. 1.14: An illustration of VSB used in TV transmission

VSB modulation is used to transmit TV picture signals whereas wideband FM is used to transmit sound signals.

Generation

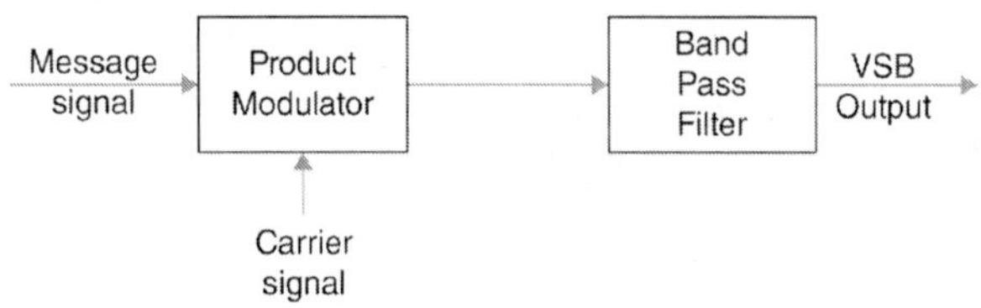

Fig. 1.15: Generation of the VSB

The product modulator produces DSB-SC signal as shown in Fig. 1.15. The bandpass filter allows only one sideband and a portion of another sideband, which is nothing but a VSB modulated signal. The coherent demodulator is used to demodulate VSB signals at the receiver.

1.8 AM TRANSMITTER

Based on the power level, the AM transmitters are classified into two types:

(i) *Low-Level AM Transmitter*

The block diagram of low-level AM transmitter is shown in Fig. 1.16. In this transmitter, the carrier signal is modulated at low-power level and then the signal is amplified.

The master oscillator produces stable oscillations. It should not be affected by the changes in temperature, supply voltage and other environmental conditions. The output of the master oscillator is given to the buffer amplifier which would load the master oscillator and also draw current from it. Next, the class-C tuned amplifier which produces multiple of carrier frequencies is used as a harmonic generator circuit.

The modulating signal is amplified by the audio amplifier and fed to the modulator power amplifier to boost the power level. The output of this amplifier and harmonic generator are given as inputs to the class-C modulated amplifier.

The modulation process is performed by this amplifier. The modulated signal is fed to the class-B linear modulated amplifier to raise the carrier power. Finally, the amplified signal is sent to the antenna to convert electrical signal to the electromagnetic signal.

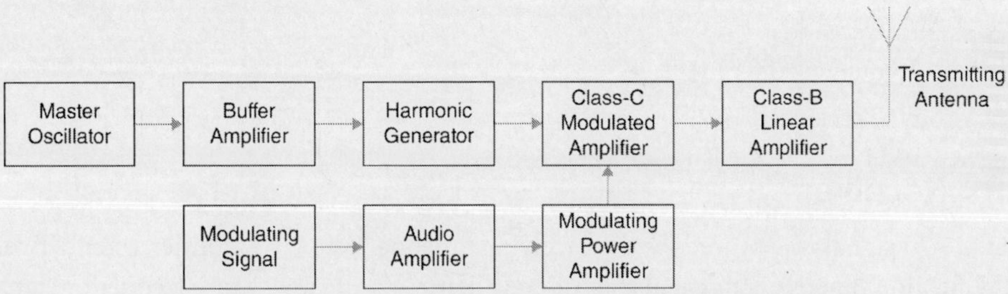

Fig. 1.16: Block diagram of a low-level AM transmitter

(ii) *High-Level AM Transmitter*

The block diagram of a high-level AM transmitter is shown in Fig. 1.17. In this transmitter, the modulating signal is first amplified and then modulated in high power level.

The master oscillator produces carrier signal. Usually, carrier frequency generated by the oscillator is stable. However, it should not be affected by temperature and ageing of the components.

The master oscillator is connected to the buffer amplifier. If the master oscillator is directly connected to the harmonic generator, then it may load the master oscillator. To avoid this buffer amplifier is placed in between to isolate them. Next, the class-C amplifier is added. The purpose of it is to boost the RF signal to higher level. Cascaded class-C amplifiers are useful to obtain higher efficiency.

On the other hand, the modulating signal is fed to the audio amplifier. It boosts the low-power modulating signal.

The Class-B push–pull amplifier acts as the modulating amplifier. It feeds the audio power into the modulated amplifier, which also receives the carrier signal from the class 'C' power amplifier.

The modulated amplifier is usually class 'c' tuned amplifier usually of push-pull type.

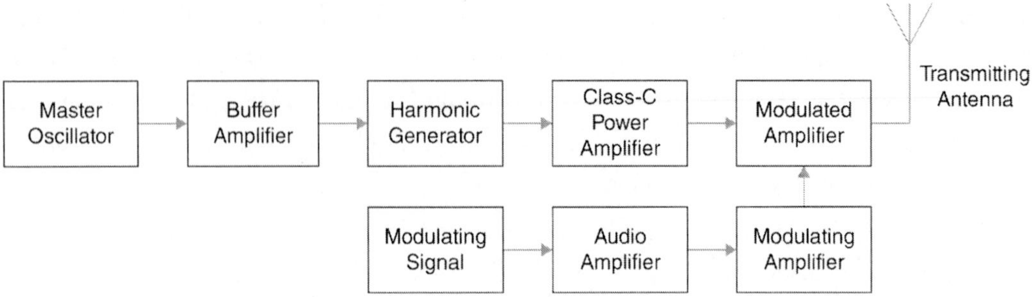

Fig. 1.17: Block diagram of a high-level AM transmitter

1.9 AM RECEIVERS

AM receiver captures the AM modulated signal and demodulate it to obtain the message signal. There are two types of AM receivers:

(i) Tuned radio-frequency (TRF) receiver

(ii) Superheterodyne receiver

1.9.1 TRF Receiver

The modulated AM signal is captured by the receiving antenna. The RF amplifier usually two or three stages of amplification tuned to a particular frequency hence it will allow those frequency signals to the next audio detector. It demodulates the AM signal and pass it to the audio amplifier where signal power is boosted. Finally, the audio signals are heard, with the help of speaker.

Advantages of the TRF receiver

(i) The circuit of TRF receiver is simple and low complex

Disadvantages

(i) The multistage RF amplifier provides instability and non-uniform gain for the broadband signals. It also fails to reject adjacent frequency signals properly.

1.9.2 Superheterodyne Receiver

The problems of TRF receiver such as instability, insufficient adjacent frequency rejection is overcome in superheterodyne receiver. The block diagram of a superheterodyne receiver is shown in Fig. 1.18.

Antenna

The antenna converts EM waves into electric signals, a tuning circuit is placed at the input of the RF amplifier to tune to the desired frequency and reject all other frequencies.

RF Amplifier

The purpose of this RF amplifier is to improve the sensitivity and selectivity of the receiver. It amplifies input RF signal and boost its signal to noise ratio.

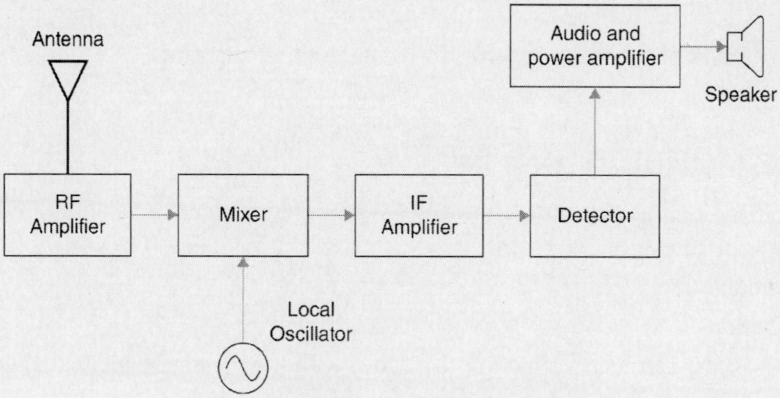

Fig. 1.18: Superheterodyne receiver

Mixer

The mixer has two circuits. (i) analog multiplier (ii) band-pass filter. The analog multiplier receives two input frequency signals, one from local oscillator f_{LO} and another from RF amplifier f_{RF}. It produces sum $(f_{LO} + f_{RF})$ and difference $(f_{LO} - f_{RF})$ or $(f_{RF} + f_{LO})$ of frequencies. The band-pass filter allows only one desired frequency called Intermediate Frequency (IF).

IF Amplifier

This amplifier provides required gain to the IF stage. It is usually a tuned amplifier with two or three stages of amplification

Detector

It is usually a demodulator circuit and reconstructs the original message signal.

AF and the Power Amplifier

It amplifies the message signal to a higher power level and sends the signal to the loud speaker.

Loud Speaker

It converts electrical signal to acoustic signal through which original message is heard.

Advantages

(i) Selectivity and sensitivity are better than TRF receivers.
(ii) Gain is high and uniform over wideband frequencies.

Sensitivity

Sensitivity is the ability of the radio receiver to respond to the weakest signals. It is measured in microvolts (μV) or decibels (dB).

Selectivity

Selectivity is the ability of the radio receiver to pick up a desired band and reject other adjacent frequencies

Fidelity

Fidelity is the ability of the radio receiver to reconstruct all the range of modulating frequencies equally.

1.10 ANGLE MODULATION

The angle of the carrier is varied in accordance with the modulating signal is called as angle modulation. Angle modulation is divided into two types:

(i) Frequency modulation

(ii) Phase modulation

The frequency of the carrier is varied in accordance with the message signal is called as frequency modulation.

The frequency modulated wave, as shown in Fig. 1.19, is expressed as,

$$S_{FM}(t) = A_c \cos\left(2\pi f_c t + 2\pi k_f \int_0^t m(\tau)\,d\tau\right) \tag{1.40}$$

where

A_c → amplitude of the modulated signal

f_c → carrier frequency

k_f → sensitivity factor

$m(\tau)$ → modulating signal

The modulating signal is represented as

$$m(t) = A_m \cos(2\pi f_m t) \tag{1.41}$$

where

A_m → amplitude of the modulating signal

f_m → frequency of the modulating signal

1.10.1 Modulation Index

The modulation index of FM is directly proportional to the peak frequency deviation (Δf) and inversely proportional to the modulating frequency (f_m).

$$\beta = \frac{\Delta f}{f_m} \tag{1.42}$$

where

$$\Delta f = k_f A_m.$$

Deviation ratio is the ratio of the peak frequency deviation (Δf) and inversely proportional to the maximum of modulating frequency ($f_{m(max)}$).

$$D = \frac{\Delta f}{f_{m(max)}} \tag{1.43}$$

The average power of the FM signal can be found from Eq. (1.42), which is given by

$$P_t = \frac{A_c^2}{2} \tag{1.44}$$

From Eq. (1.40), FM wave is given by

$$S_{FM}(t) = A_c \cos\left(2\pi f_c t + 2\pi k_f \int_0^t m(\tau)\,d\tau\right).$$

Substituting Eq. (1.41) in (1.42) results in,

$$\begin{aligned} S_{FM}(t) &= A_c \cos\left(2\pi f_c t + 2\pi k_f \int_0^t A_m \cos(2\pi f_m \tau)\,d\tau\right) \\ &= A_c \cos\left(2\pi f_c t + \frac{k_f A_m}{f_m}\sin(2\pi f_m t)\right) \\ &= A_c \cos\left(2\pi f_c t + \frac{2\pi k_f A_m}{2\pi f_m}\left[\sin(2\pi f_m \tau)\right]_0^t\right) \\ &= A_c \cos(2\pi f_c t + \beta \sin(2\pi f_m t)). \end{aligned} \tag{1.45}$$

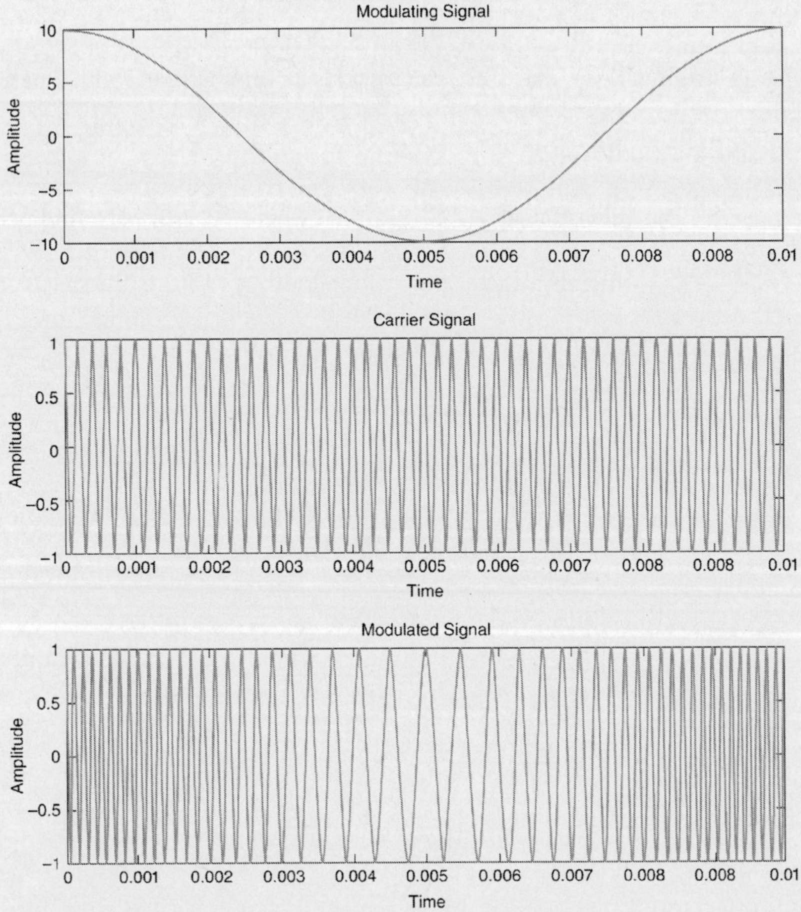

Fig. 1.19: FM waveform

Based on the modulation index β, FM is classified into narrowband FM and wideband FM. If $\beta << 1$, it is termed as narrowband FM; otherwise wideband FM.

1.10.2 Spectrum of FM

The FM wave in Eq. (1.45) can be represented as

$$S_{FM}(t) = Re\left\{e^{j(2\pi f_c t + \beta \sin(2\pi f_m t))}\right\} \tag{1.46}$$

where $Re \rightarrow$ real value of the term. For example, $z = x + ij$ with $Re(Z) = x$, $Im(Z) = y$, and remember $e^{j\theta} = \cos\theta + j\sin\theta$. The expression in Eq. (1.46) can be rewritten as

$$S_{FM}(t) = Re\left\{A_c e^{j(\beta \sin(2\pi f_m t))} e^{j(2\pi f_c t)}\right\} \tag{1.47}$$

Let us consider

$$S_b(t) = A_c\left\{e^{j(\beta \sin(2\pi f_m t))}\right\} \tag{1.48}$$

$$S_{FM}(t) = Re\left\{S_b(t) e^{j(2\pi f_c t)}\right\}. \tag{1.49}$$

Let us find the spectrum of $S_b(t)$ first. Since $\sin(2\pi f_m t)$ is a periodic signal with a period of $T = \frac{1}{f_m}$.

Thus, $e^{j(\beta \sin(2\pi f_m t))}$ is also a periodic signal.

That is, $A_c e^{j\left(\beta \sin\left(2\pi f_m\left(t + \frac{k}{f_m}\right)\right)\right)}$ can be written as $A_c e^{j(\beta \sin(2\pi f_m t + 2k\pi))}$, where $k = 1, 2, \ldots$ to find periodicity. This is equivalent to $A_c e^{j(\beta \sin(2\pi f_m t))}$, which is also periodic with period $\frac{1}{f_m}$. Hence, the spectrum of $S_b(t)$ can be found using the discrete Fourier Series, as given below,

$$S_b(t) = \sum_{n=-\infty}^{\infty} C_n e^{j(2\pi n f_m t)} \tag{1.50}$$

where

$$C_n = \frac{1}{T}\int_0^T S_b(t) e^{-j(2\pi n f_m t)} dt$$

$$C_n = \frac{1}{T}\int_0^T A_c e^{j\beta \sin(2\pi f_m t)} e^{-j2\pi n f_m t}\, dt$$

$$= \frac{1}{T}\int_0^T A_c e^{j(\beta \sin(2\pi f_m t - 2\pi n f_m t)}\, dt. \tag{1.51}$$

Put $2\pi f_m t = x$, $dt \quad \frac{dx}{\quad}$

Lower limit: $x = 2\pi f_m t = 0$

Upper limit: $x = 2\pi f_m t = 2\pi \frac{1}{T} \times T$

$$x = 2\pi \tag{1.52}$$

Substituting Eq. (1.51) and (1.52) in (1.50), we get

$$C_n = \frac{1}{T}\int_0^{2\pi} A_c e^{j(\beta\sin(x)-nx)} \frac{dx}{2\pi f_m}$$

$$C_n = \frac{A_c}{2\pi}\int_0^{2\pi} e^{j(\beta\sin(x)-nx)} dx \tag{1.53}$$

Define the Bessel function of first kind order 'n'

$$J_n(\beta) = \frac{1}{2\pi}\int_0^{2\pi} e^{j(\beta\sin(x)-nx)} dx \tag{1.54}$$

After substituting Eq. (1.54) in (1.53), it becomes

$$C_n = A_c J_n(\beta) \tag{1.55}$$

Substituting Eq. (1.55) in (1.50) results in

$$S_b(t) = \sum_{n=-\infty}^{\infty} A_c J_n(\beta) e^{j2\pi n f_m t} \tag{1.56}$$

Substituting (1.56) in (1.49)

$$S(t) = \sum_{n=-\infty}^{\infty} \mathrm{Re}\left[A_c J_n(\beta) e^{j2\pi n f_m t} e^{j2\pi f_c t}\right]$$

$$S(t) = \sum_{n=-\infty}^{\infty} \mathrm{Re}\left[A_c J_n(\beta) e^{j2\pi(nf_m+f_c)t}\right] \tag{1.57}$$

Simplifying Eq. (1.57) results in

$$S_{FM}(t) = \sum_{n=-\infty}^{\infty} A_c J_n(\beta)\cos\left(2\pi\left(f_c+nf_m\right)t\right) \tag{1.58}$$

To find the spectrum, the below trigonometric formula is applied,

$$\cos\left(2\pi f_0 t\right) \xrightarrow{\text{FT}} \frac{1}{2}\delta\left(f-f_0\right)+\frac{1}{2}\delta\left(f+f_0\right) \tag{1.59}$$

Similarly,

$$\cos\left(2\pi(f_c+nf_m)t\right) \xrightarrow{\text{FT}} \frac{1}{2}\delta\left(f-f_c-nf_m\right)+\frac{1}{2}\delta\left(f+f_c+nf_m\right) \tag{1.60}$$

Using Eq. (1.60), we can find the spectrum of FM

$$S_{FM}(f) = \sum_{n=-\infty}^{\infty}\frac{1}{2}A_c J_n(\beta)\left[\left(f-f_c-nf_m\right)+\left(f+f_c+nf_m\right)\right] \tag{1.61}$$

Although $n \in \{-\infty, \infty\}$ (implies bandwidth is infinity), a realistic value of bandwidth of FM signal is those spectral components ($f_c + nf_m$) for which $J_n(\beta)$ is significant,

$$\lim_{n\to\infty} J_n(\beta) = 0 \tag{1.62}$$

1.10.3 Power

The total power of the modulating signal is distributed among carrier component and sideband components. The average power of modulated carrier signal is given as

$$P_t = \frac{A_c^2}{2} \tag{1.63}$$

1.10.4 Carson's Bandwidth Formula

Carson has given an approximate formula for the bandwidth of an FM signal, which is given by

$$B = 2(\Delta f + f_m) \tag{1.64}$$

1.11 GENERATION OF THE FM WAVE

In FM, the amplitude changes of the modulating signal is translated into frequency changes of the carrier. The carrier frequency can be generated by LC oscillators. The diodes such as varactor diode can act as a capacitor in the reversed biased condition. There are two circuits elements, L_1, and C_1, as shown in Fig. 1.20, which form a tank circuit to generate the FM signal. The capacitance of the varactor diode is varied by the modulating signal and the pair of resistors R_1 and R_2.

The modulating signal (AF) is fed through the capacitors. The radio-frequency choke (RFC) function is to provide isolation of the carrier frequency tank circuit and the AF input.

During the positive half-cycle of the modulating signal, the total negative voltage in tank circuit increases because of positive half-cycle and reverse bias of the varactor diode. It decreases the capacitance, which in turn increases carrier frequency. During negative half-cycle, it reduces the reverse bias of the varactor diode, which in turn increases the capacitance. This reduces the carrier frequency.

Therefore, the carrier frequency produced by the tank circuit varies with respect to the modulating signal, which produces desired FM signal.

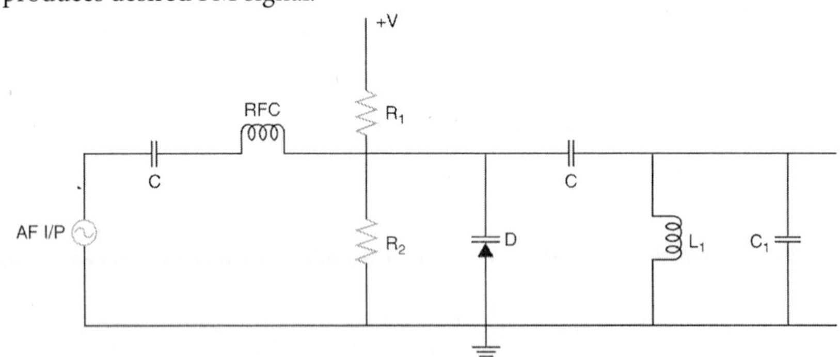

Fig. 1.20: FM generation by the varactor diode method

1.12 FM DETECTION USING THE PLL

Phase-locked loop (PLL) is a popular FM demodulation technique. PLL is basically consisting of a multiplier, a loop filter and a voltage-controlled oscillator as shown in Fig. 1.21. FM detection using PLL is a negative feedback system.

A voltage-controlled oscillator (VCO) is a device which produces oscillations depending upon applied input voltage. When input voltage varies, the output frequency also changes. Therefore, it can act as a FM modulator when modulating signal applied as an input signal.

The primary function of the PLL is to track the phase of the carrier signal. PLL is a simple reliable FM demodulation circuit to detect FM signal with low power and much noise.

In PLL, the feedback signal tends to follow the input signal. If the signal fed from VCO is not equal to the incoming FM signal, an error signal is produced by the multiplier. The error signal is nothing but the difference between $s(t)$ and $f(t)$. The error signal is passed through loop filter to remove high frequency components.

The detected output signal is the demodulated signal which is the original modulating signal. This signal $v(t)$ is fed as the input to the VCO to produce the locally generated FM signal $f(t)$.

When the phase of the feedback signal $f(t)$ is equal to the incoming FM signal $s(t)$ then the error signal $e(t)$ becomes zero. This is the time, PLL is locked and produces required modulating signal.

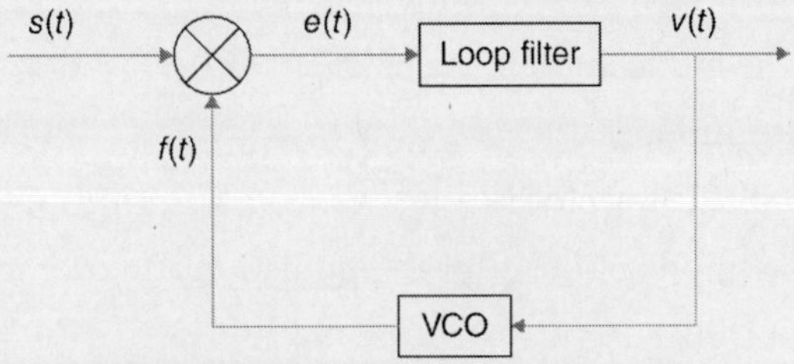

Fig. 1.21: FM detection using PLL

UNIT SUMMARY

In this unit, two important modulation techniques (AM and FM) have been studied. Both AM and FM belongs to continuous wave modulation.

Amplitude Modulation

- The variations of the message signal are translated to amplitude variations of the carrier signal.
- AM is classified into DSB-FC, DSB-SC, SSB-SC and VSB.
- DSB-FC wave is transmitted with both sidebands with full carrier power. The modulation index is a function of peak frequency deviation and modulating frequency. Envelope detector circuit is used to demodulate the modulated signal.
- DSB-SC wave is transmitted with both sidebands with suppressed carrier power. Coherent detector is used to demodulate the DSB-SC signal.
- The bandwidth DSB-FC or DSB-SC is twice the bandwidth of the message signal.
- SSB-SC is generated by frequency and phase discrimination methods. It is a spectral and power efficient technique because it occupies only half the bandwidth of DSB-SC technique.

- VSB is used in the TV picture carrier transmission. The bandwidth of VSB is one full sideband and a portion of another sideband.

Frequency Modulation

- The variations of the message signal are translated to frequency variations of the carrier signal.
- It is inferred from the spectral analysis that if the modulation index is lesser than one (exactly 0.3) then it is called narrowband FM and otherwise it is wideband FM.
- The bandwidth of the FM is approximated by CARSON rule which is given as $B = 2(\Delta f + f_m)$.
- The FM is generated through varactor diode circuit and the detection is done through PLL circuit.

EXERCISES

Two mark questions

1. Define modulation.
2. Brief the need for modulation in terms of antenna height.
3. What is the formula for the modulation index of DSB-FC?
4. Classify the modulation techniques.
5. How do you represent DSB-FC waveform?
6. How do you classify under modulation, critical modulation and over modulation?
7. What is the bandwidth of DSB-FC?
8. How do you relate the carrier power and total power of an AM signal?
9. How much percentage of power lies in the sideband of DSB-FC waveform?
10. Which is the correct choice of time constant value in an envelope detector circuit?
11. Why envelope detector cannot be used for DSB-SC demodulation?
12. Mention the meaning of 'coherent' in coherent modulation.
13. What is the purpose of LPF in a coherent detector?
14. Write down two different methods of generation of SSB-SC.
15. What is the application of VSB modulation?
16. Define sensitivity.
17. Define selectivity.
18. Define fidelity.
19. Write the bandwidth of FM signal using Carson's rule.
20. What are the blocks of PLL?

Five mark questions

1. Discuss elaborately the need for modulation.
2. Derive the spectrum (frequency domain) of DSB-FC signal with spectrum diagram.

3. Explain the operation of an envelope detector with a neat diagram and necessary conditions.
4. Derive the total power transmitted by AM signal.
5. Explain the working of a coherent detector with necessary equations.
6. Derive the frequency domain representation of a wideband FM signal.
7. Explain the generation of FM signal using a varactor diode circuit.
8. Describe the working of phase-locked loop with a neat diagram.

Ten mark questions

1. Explain the two generation methods of SSB-SC modulation with neat block diagrams.
2. How do you broadcast the AM wave using low-power and high-power transmitters? Narrate it in detail.
3. Discuss in detail the working of a superheterodyne receiver with a neat diagram.

Numerical Problems

1. The total sideband power for an AM signal is given as 100 W with 50% of modulation. Find the total power of transmitted AM signal.

Solution

The sideband power is given by P_{SB} = 100 W, the modulation index (μ) is given as 50%, i.e., μ = 0.5

$$P_{SB} = \frac{P_C \mu^2}{2},$$

where P_C is the carrier power

$$P_C = \frac{2P_{SB}}{\mu^2}$$

$$P_C = \frac{2 \times 100}{0.5^2}$$

$$P_C = \frac{200}{0.25}$$

$$P_C = 100 \text{ W.}$$

Therefore,

$$P_T = P_C + P_{SB}$$

$$P_T = 800 + 100$$

$$P_T = 900 \text{ W.}$$

The total AM transmitted power is 900 W.

2. The carrier and each of the sideband power of an AM signal are 8 kW and 2 kW, respectively. What is the percentage of modulation?

Solution

Given,

Carrier power P_C = 8 kW

Sideband power = P_{SB}

P_{SB} = lower sideband power + upper sideband power

$$= P_{USB} + P_{LSB}$$

$$= 2 \text{ kW} + 2 \text{ kW}$$

$$P_{SB} = 4 \text{ kW}$$

The sideband power can be written as

$$P_{SB} = \frac{P_C \mu^2}{2}$$

$$\mu^2 = \frac{2P_{SB}}{P_C}$$

$$\mu^2 = \frac{2 \times 4}{8}$$

$$\mu = 1.$$

The percentage of modulation = $\mu \times 100\% = 100\%$

3. What is the modulation index of the AM signal, if the peak voltage varies from 2 V to 10 V?

Solution

Given,

Maximum amplitude, $A_{max} = 10$ V

Minimum amplitude, $A_{min} = 2$ V

The modulation index, $\mu = \frac{A_{max} - A_{min}}{A_{max} + A_{min}}$

$$= \frac{10-2}{10+2}$$

$$= \frac{8}{12}$$

Modulation index = 0.66

4. An AM signal is given by $S(t) = A_c \cos \omega_c t + 2\cos \omega_c t \cos \omega_m t$. What should be the minimum amplitude of carrier signal (A_c) for distortionless demodulation using an envelop detector?

Solution

Given,

Comparing with AM signal S(t) with S_{AM} (t)

$$S(t) = A_c \left\{1 + \frac{2}{A_c} \cos \omega_m t\right\} \cos 2\pi f_c t$$

$$S_{AM}(t) = A_c \{1 + \mu \cos \omega_m t\} \cos 2\pi f_c t$$

$$\mu = \frac{2}{A_c}.$$

For envelope detection, $\mu \le 1$

$$\frac{2}{A_c} \le 1$$

$$A_c \ge 2$$

Hence, the minimum value of A_c is 2 V.

5. The carrier power of an AM signal is given by 500 W. Find the amount of power saved if carrier and one of the sidebands is suppressed with modulation index, $\mu = 0.5$?

Solution

Given,

$$P_C = 500 \text{ W}$$

$$P_T = P_C\left\{1 + \frac{\mu^2}{2}\right\}$$

$$= 500\left\{1 + \frac{0.25}{2}\right\}$$

$$= 562.5 \text{ W.}$$

So, the percentage of power saved in SSB $= \dfrac{4+\mu^2}{4+2\mu^2}$

$$= \frac{4+0.5^2}{4+2(0.5)^2}$$

Percentage of power saved = 94.5%

Amount of power saved = 94.5% of 501.125

$$= \frac{94.5}{100} \times 501.125$$

Amount of power saved = 531.56 W.

6. An unmodulated carrier frequency is given by 1 MHz. The maximum frequency after frequency modulation is given by 1.4 MHz. Find the frequency deviation (Δf) and minimum frequency (f_{min}) of an FM signal.

Solution:

Given,

Carrier frequency, $f_c = 1$ MHz

Maximum frequency, $f_{max} = 1.4$ MHz

$$= f_c + \Delta f$$

$$1.4 \text{ MHz} = 1 + \Delta f$$

$$\Delta f = 0.4 \text{ MHz}$$

and $f_{min} = f_c - \Delta f$

$$= 1 - 0.4$$

$$f_{min} = 0.6 \text{ MHz.}$$

The frequency deviation and the minimum frequency of a FM signal are 0.4 MHz and 0.6 MHz, respectively.

7. The maximum frequency (f_{max}) of an FM signal is given by 1.5 MHz. The total frequency swing is given by 900 kHz. What is the frequency deviation and minimum frequency?

Solution:

$$\text{Total frequency swing} = 2\Delta f$$

$$= 900 \text{ kHz}$$

$$\Delta f = 450 \text{ kHz}$$

Now,

$$f_{max} = 1.5 \text{ MHz}$$

$$= f_c + \Delta f$$

$$f_c = f_{max} - \Delta f$$

$$= 1.5 \times 1000 - 450$$

$$= 1050 \text{ kHz}$$

$$f_c = 1.05 \text{ MHz}$$

$$f_{min} = f_c - \Delta f$$

$$= 1.05 \times 1000 - 450$$

$$f_{min} = 600 \text{ kHz.}$$

The frequency deviation is 450 kHz and the minimum frequency is 600 kHz.

8. A sinusoidal carrier of 20 V, 2 MHz is frequency modulated by a message signal of $10\sin\left(4\pi 10^3 t\right)$. k_f is given by 50 kHz/V. Find f_{max} of the FM signal.

Solution:

Given,

$$c(t) = A_c \cos 2\pi f_c t$$

So,

$$A_c = 20 \text{ V}, f_c = 2 \quad \text{MHz}$$

$$m(t) = A_m \sin\left(4\pi 10^3 t\right)$$

$$A_m = 10, f_m = 2 \text{ kHz}$$

$$k_f = 50 \text{ kHz / V}$$

So,

$$f_{max} = f_c + k_f \left|m(t)\right|$$

$$= 2000 + 50 \times 10$$

$$f_{max} = 2500 \text{ kHz.}$$

The maximum frequency of FM signal is 2500 kHz.

9. A carrier is frequency modulated with maximum frequency deviation of 16 kHz with message signal frequency given by 4 kHz. Find the modulation index (β) and bandwidth of the FM signal.

Solution:

Given,

$$\Delta f = 16 \text{ kHz}$$

$$f_m = 4 \text{ kHz}$$

Now,

$$\Delta f = \beta f_m$$

$$\beta = \frac{\Delta f}{f_m}$$

$$\beta = 4$$

$$\text{Bandwidth} = 2(\beta + 1) f_m$$

$$= 2 \times 5 \times 4 \text{ kHz}$$

$$\text{Bandwidth} = 40 \text{ kHz}$$

The modulation index and bandwidth of FM signal are 4 and 40 kHz, respectively.

10. A carrier is frequency modulated with maximum frequency deviation of 100 kHz. Find the modulation index (β) and bandwidth if the message signal frequency is 10 kHz.

Solution:

Given,

$$\Delta f = 100 \text{ kHz}$$

$$f_m = 10 \text{ kHz}$$

Now,

$$\Delta f = \beta f_m$$

$$\beta = \frac{\Delta f}{f_m}$$

Modulation index, $\beta = 10$

So, bandwidth $= 2(\beta + 1) f_m$

$$= 2 \times 11 \times 10 \text{ kHz}$$

$$\text{Bandwidth} = 220 \text{ kHz}$$

The modulation index and bandwidth of FM signal are 10 and 220 kHz, respectively.

KNOW MORE

- It is incredible that radio waves were predicted before they were ever found. James Clark Maxwell predicted the presence of electromagnetic waves.
- 'Physicist Heinrich Hertz from Germany Italian Guglielmo Marconi discovered the existence of electromagnetic waves in 1886, and a decade later he had created a useful radio transmission and reception equipment.'

- 'From Ocean Bluff-Brant Rock, Massachusetts, Reginald Fessenden sent a message to ships at sea in 1906. O Holy Night performed on the violin was the subject of the programme. It was the first occasion AM, or what is now known as AM radio, had been used.'
- During the radio industry's Golden Age, AM was the dominant technology. This was the majority of families' main source of news and amusement prior to television.
- American Edwin Armstrong invented a novel way to transmit audio using radio waves in 1933. The term 'frequency modulation' applied to this. It would change the signal's frequency rather than loudness. This turned out to be a much better method of sound transmission than AM. Additionally, it permitted the multiplexing of signals, enabling the stereo broadcast of music.

Guglielmo Marconi. Italy's Bologna is where Marconi was born. His mother was Irish, while his father was Italian. Bologna was his first place of study, followed by Florence. After that, he moved to Leghorn's technical college to study physics. Italian inventor Guglielmo Marconi constructed the apparatus in 1895 and used it to carry electrical signals from one end of his home to the other and then from the home to the garden. In actuality, these trials marked the beginning of widespread wireless radio or telegraphy. Marconi became fascinated on the concept of sending messages over the Atlantic after the success of his experiments at home. At Poldhu, on the southwest tip of England, he constructed a transmitter that was 100 times more powerful than any previous station, and at St. John's, Newfoundland, he set up a receiving station in November 1901. He got signals from across the ocean on 12 December 1901. As word of this achievement travelled throughout the world, eminent scientists like Thomas Edison praised him. Numerous awards were given to Marconi, including the Nobel Prize in Physics in 1909. He was assigned as a representative to the Paris Peace Conference in 1919, where he participated in the signature of the peace agreements with Austria and Bulgaria.

Edwin Howard Armstrong. The first trans-Atlantic radio communication was made by Guglielmo Marconi while Edwin Howard Armstrong was only 11 years old. Armstrong was interested with radio that he started searching and studying it and creating his own wireless devices, which includes the 125-foot antenna established in his parent's place. Armstrong attended Columbia University because of his passion in science and technology, where he studied at the school's Hartley Laboratories and left a lasting impression on a number of his teachers. He earned an electrical engineering degree from college when he graduated in 1913. The same year he obtained his diploma and he invented the regenerative circuit. By running the signal through a radio tube over 20,000 times/sec, regeneration amplification boosted the strength of the radio signal that was received and increased the range of radio broadcasts. Armstrong received a patent for this creation in 1914. Armstrong originally looked into the issue because AM radio has previously been incredibly vulnerable to this type of interference. He carried out his research in the Philosophy Hall basement of Columbia University. Armstrong's FM technique was granted a U.S. patent in 1933. As FM technology developed throughout the 1930s, it became increasingly competitive with already-existing technologies. The Federal Communications Commission (FCC) agreed to establish a commercial FM service in 1940 and the service came online the following year with 40 channels. Nevertheless, the start of World War II constrained the amount of funding that could be allocated to new radio infrastructure.

REFERENCES AND SUGGESTED READINGS

1. Simon Haykin: *Communication Systems*, John Wiley & Sons, New York, Fourth Edition, 2001.
2. Taub and Schilling: *Principles of Communication Systems*, Tata McGraw-Hill, New Delhi, 1995.

References for further reading

1. https://nptel.ac.in/courses/108104091
2. https://onlinecourses.nptel.ac.in/noc23_ee117/preview

2 Pulse Analog Modulation

UNIT SPECIFICS

In this unit, we will be discussing the following aspects:

- Theory of ideal sampling
- Definition of sampling theorem
- Aliasing phenomenon
- Concept of interpolation
- Pulse amplitude modulation technique
- Natural and flat top sampling in time and frequency domains

RATIONALE

This unit on pulse amplitude modulation begins with the description of the process of ideal sampling and frequency domain representation of the sampled signals. The definition of sampling theorem is stated in detail. Subsequently, if sampling process is not performed as per sampling theorem then how aliasing arises have been illustrated elaborately. The techniques to overcome aliasing are also suggested. The concept of interpolation is described in detail. The pulse amplitude modulation is elaborated with help of sample-and-hold circuit. Finally, the significance of natural and flat top sampling process is emphasized with neat diagrams.

PRE-REQUISITES

Class 10 science

UNIT OUTCOMES

List of outcomes of this unit is as follows:

U2-O1: Understand the fundamental theory of ideal sampling process.

U2-O2: Define the sampling theorem.

U2-O3: Discuss the concept of interpolation phenomenon.

U2-O4: Study the pulse amplitude modulation technique.

U2-O5: Analyse the flat top and natural sampling in time and frequency domains.

Unit-2 Outcome	EXPECTED MAPPING WITH COURSE OUTCOMES (1 - Weak Correlation; 2 - Medium Correlation; 3 - Strong Correlation)					
	CO-1	CO-2	CO-3	CO-4	CO-5	CO-6
U2-O1	-	3	-	1	-	2
U2-O2	-	3	-	1	-	2
U2-O3	-	3	-	1	-	-
U2-O4	-	3	-	1	-	3
U2-O5	-	3	-	1	-	3

MODULE 2: PULSE ANALOG MODULATION

2.1 IDEAL SAMPLING

In an ideal sampling, the signal that is used to sample the input signal say $x(t)$ is an impulse train. This approach is also called instantaneous or impulse train sampling. In a perfect sampling method, the instantaneous value of $x(t)$ is equal to the area of each impulse in the sampled signal. Fig. 2.1 is an example of ideal sampling.

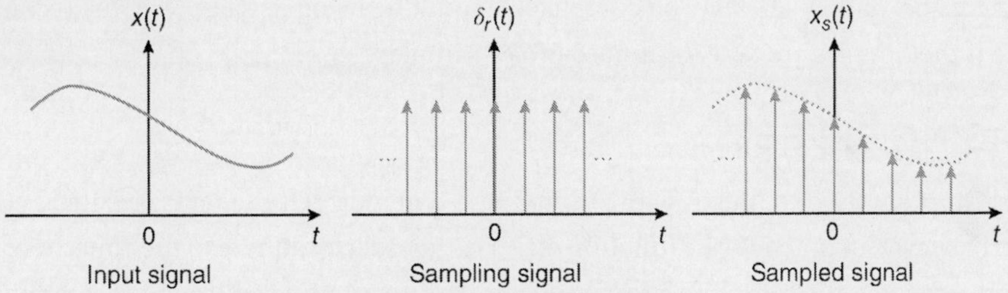

Fig. 2.1: Ideal sampling

2.2 FREQUENCY–DOMAIN DESCRIPTION OF SAMPLING

Let us consider a stochastic process $g(t)$ that exhibits random behaviour and has limited energy, defined for all values of time t.

(1) Analog signal

(ii) Analog signal (Sampled version)

The analog signal $g(t)$ is shown in part in Fig. 2.2(a). An infinite sequence of samples is created when the signal $g(t)$ is sampled at a constant rate as shown in Fig. 2.2(b), for example, once per T_s seconds. These samples may be expressed as $\{g(nT_s)\}$, where n is the total number of potential values, positive and negative and are spaced T_s apart. In this context, the term 'T_s' may be used to denote the sample period, whilst it's reciprocal, denoted as '$fs = 1/T_s$,' can be referred to as the *sampling rate*. Therefore, due to these factors, the *sampling method* in question is referred to as *instantaneous sampling*. It is crucial to remember that a uniform sampling of a continuous-time signal with limited energy will result in a spectrum that is periodic, with the sampling rate equal to the frequency of the periodic components.

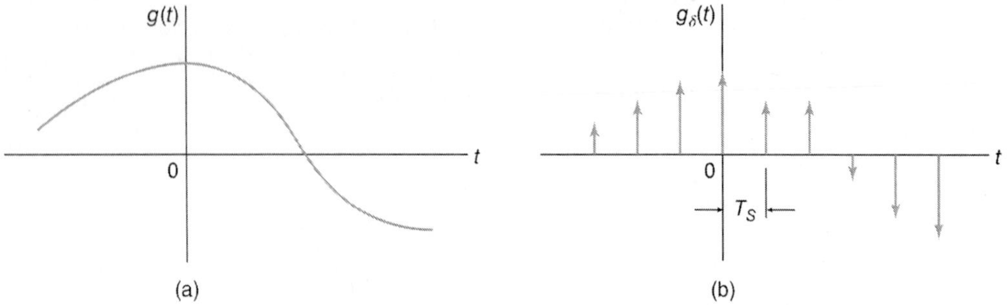

Fig. 2.2: (a) analog signal and (b) sampled analog signal

2.3 THE SAMPLING THEOREM

The sampling theorem may be expressed as follows when applied to strictly band-limited signals with finite energy:

(i) A signal that has a finite amount of energy and is restricted to a certain range of frequencies, with no components exceeding W Hz, may be completely characterised by the values of the signal at specific time instances that are separated by intervals of $1/2\ W$.

(ii) A signal that has finite energy and is band-limited, meaning it does not include any frequency components beyond W Hz, may be completely rebuilt by using its samples. This reconstruction is possible if the samples are taken at a pace of $2W$ samples per second.

2.4 ALIASING EFFECT

A common assumption used to illustrate the sampling theorem is that the message signal $g(t)$ is strictly band-limited. However, in practice, the function $g(t)$ does not strictly adhere to the concept of being band restricted. Consequently, the sampling process encounters a certain level of under sampling, resulting in *aliasing*. A signal's spectrum exhibits 'aliasing' when a high-frequency component seems to be a low-frequency component of its sampled counterpart. Fig. 2.3 illustrates this impact.

The spectrum that exhibits aliasing, as seen in Fig. 2.3(b) (represented by a solid curve), corresponds to the signal of the message that has been under sampled, as displayed in Fig. 2.3(a).

To overcome the effects of aliasing the following methods can be used:

(i) Low-pass anti-aliasing filters are often applied prior to sampling in order to minimise the quantity of high frequency components in the signal that are irrelevant to the data that the message signal is delivering.

(ii) To ensure accurate signal reconstruction, it is recommended to sample the filtered signal at a sampling rate slightly over the Nyquist rate.

To enhance the process of designing the reconstruction filter, which aims to recover the original signal from its sampled representation, it is beneficial to use a sampling rate that exceeds the Nyquist rate.

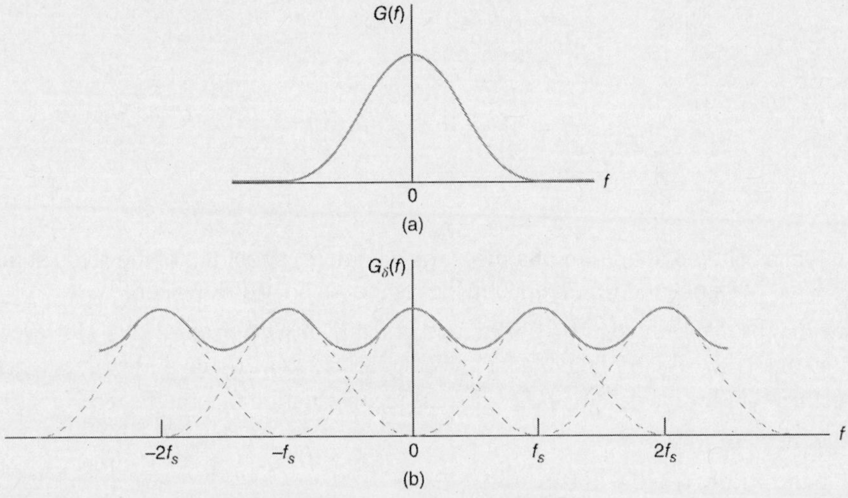

Fig. 2.3: (a) Spectrum of a signal and (b) spectrum of an under- sampled signal showing the aliasing effect.

Let us now assume, by way of illustration, the scenario in which a message signal undergoes anti-aliasing (low-pass) filtering. Fig. 2.4(a) displays the generated spectrum, whereas Fig. 2.4(b) illustrates the spectrum in relation to the immediately sampled version of the signal. It is assumed that the sample rate in this case is higher than the Nyquist rate. Thus, using the information shown in Fig. 2.4(c), the reconstruction filter's design may be stated as follows.

The anti-aliasing filter determines the passband of the reconstruction filter, which is a low-pass filter with a range of $-W$ to W. Moreover, it is important to note that the reconstruction filter is designed to have a transition band that spans from positive frequencies (W) to the difference between the sampling rate (f_s) and W, that is ($f_s - W$).

Example

As an example, let us consider the sampling of voice signals in order to perform waveform coding. For telephonic communications, the band of frequency that ranges from 100 Hz to 3.1 kHz is usually considered to be sufficient. Frequency band limiting is the process of sending the voice stream through a 3.1 kHz cut-off frequency low-pass filter. This filter is an anti-aliasing filter. Subsequently, the

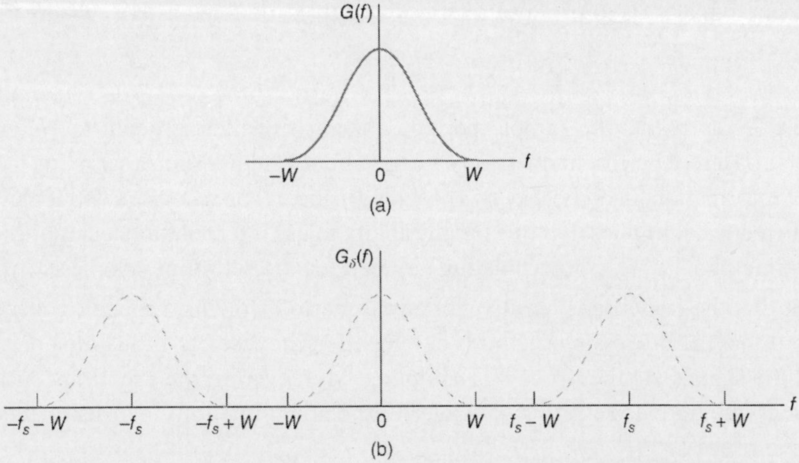

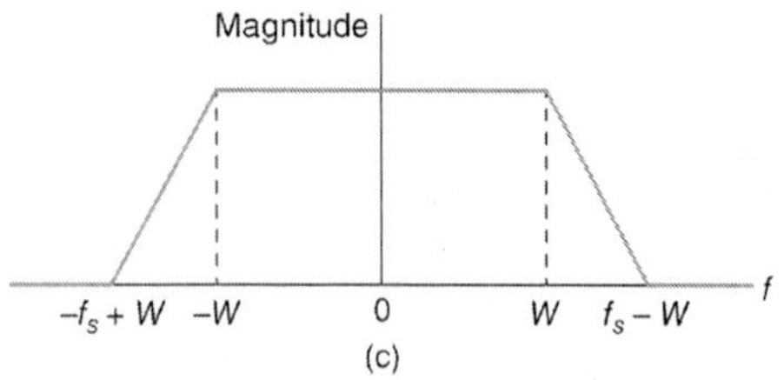

Fig. 2.4: (a) Anti- alias filtered spectrum of a message signal, (b) spectrum of the Nyquist rate sampled signal and (c) magnitude plot of the reconstruction filter

Nyquist rate may be determined as $f_s = 2 \times 3.1$ kHz, resulting in a value of 6.2 kHz. However, to perform waveform coding, the standard sampling rate for voice signals is 8 kHz. Using these values together, we can arrive at the specifications for the design of the low-pass reconstruction filter at the receiver as given below:

Cutoff frequency: 3.1 kHz

Transition band width: 6.2 to 8 kHz

Transition bandwidth: 1.8 kHz.

2.5 INTERPOLATION PHENOMENON

Interpolation is a useful signal processing that comes into play if we want to convert discrete-time signals to continuous-time signals. Specifically, this is required to interface an analog system with a digital system. As an example, let us look at a situation in which a loudspeaker and an analog amplifier are driven by a discrete-time waveform synthesiser. It will be helpful in this situation to show the amplifier's input as a function of a real variable that is specified across the whole real line. This is so because continuous-time functions provide an accurate representation of the operation of any analog circuitry.

The fundamental aspect of the interpolation process is in the correlation between the physical time length T_s and the intervals between samples within a discrete-time sequence. Moreover, it is essential that the interpolation include the spectral characteristics of the interpolated function in relation to the original sequence. In order to get more insight, let us investigate the procedure of using uniform sampling to sample a continuous-time signal. Periodic sampling is a frequently used technique for acquiring a discrete-time representation of a continuous-time signal. This involves uniformly sampling the continuous time signal $x_c(t)$ to produce a series of samples, denoted as $x(n)$, where n represents the discrete-time index.

$$x(n) = x_c(nT_s), -\infty < n < \infty.$$

The variable T_s represents the sample period, whereas f_s, which is equal to $1/T_s$, represents the sampling frequency. Here, a pertinent question emerges about whether such a sampling technique would result in a loss of information, namely, if $x(n)$ is provided, can $x_c(t)$ be reconstructed for any value of t ? If the answer is affirmative, it implies that one has the ability to use interpolation techniques to reconstruct the continuous-time signal $x_c(t)$ by interpolating between the discrete-time samples $x_c(nT_s)$.

In the event that the response is negative for any scenario involving a specific collection of signals, it may be inferred that the processing methods developed in the discrete-time domain can be extended to continuous-time signals via the process of sampling. Moreover, in the event that such a scenario is feasible, it becomes imperative to establish suitable interpolating functions in order to effectively execute the reconstruction process.

2.6 PULSE-AMPLITUDE MODULATION

Having understood the quintessence of the sampling process, the present discussion aims to provide a formal definition of PAM, which serves as the fundamental and essential type of pulse modulation. PAM may be precisely characterized as follows:

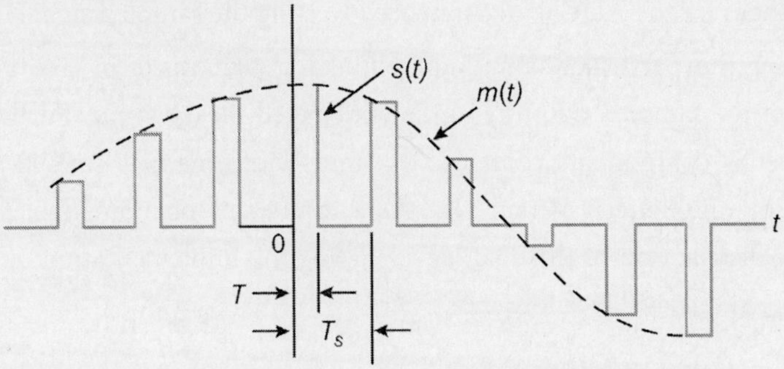

Flat-top samples, representing an analog signal.

Fig. 2.5: PAM signal

PAM is a term for a linear modulation method in which the instantaneous values of a message signal at certain sample instants are used to proportionately modify the amplitudes of regularly spaced pulses. These pulses may take the form of rectangular or other suitable shapes. Fig. 2.5 shows the waveform of a typical PAM signals.

The dashed curve in the above illustration Fig. 2.5 describes the message signal $m(t)$ that is continuous, while the rectangular pulses shown in solid lines are amplitude modulated representing the equivalent PAM signal $s(t)$.

The following two processes are carried out in the realization of the PAM signal:

(i) The sampling rate $f_s = 1/T_s$ is selected in compliance with the sampling theorem and the message signal $m(t)$ is sampled at *discrete time intervals* of T_s seconds.

(ii) To achieve a constant value T, the length of each sample is *lengthened.*

The aforementioned pair of processes are collectively referred to as 'sample and hold' within the realm of digital circuitry. The fundamental reason for purposely extending the length of each sample is to reduce the requirement for excessive channel capacity, since bandwidth and pulse duration have an inverse relationship. Also, we have some practical considerations in transmitting a PAM signal as it has stringent constraints on the channel's frequency response. This is because, the time period of the pulses that are transmitted are relatively short.

Moreover, in the context of modulation, if one were to depend on amplitude as a parameter, it is impossible for the noise performance of PAM to surpass that of baseband signal transmission. Consequently, we may use PAM as a first step towards message processing for transmission over a communication channel and then change the PAM to appropriate form of pulse modulation. In this context, and with the aim of analog-to-digital conversion in mind, we need to find the appropriate modulation that can be built on PAM which will be dealt in detail in the next module.

2.7 NATURAL AND FLAT TOP SAMPLING

Sampling may be conducted using two primary strategies, namely natural sampling and flat-top sampling. The natural sampling procedure, as seen in Fig. 2.6, involves the preservation of the original form of the sample pulses at the tops during the *sampling period.* Consequently, the process of natural sampling poses challenges for the ADC in accurately converting the sampled signal into a PCM code.

Furthermore, using this particular sampling methodology will result in dissimilarity between the frequency spectrum of the sampled output and the anticipated ideal sample. Furthermore, for higher harmonics, the frequency components produced by finite-width, narrow sample pulses will tend to decrease and will follow the pattern of (sin x)/x. As a result, this phenomenon has the potential to modify the frequency distribution of the data, necessitating the implementation of frequency domain equalization before signal retrieval by a low-pass filter.

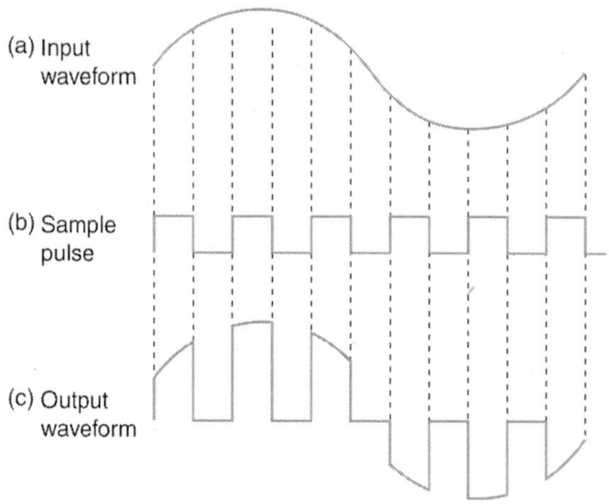

Fig. 2.6: Natural sampling: (a) analog signal, (b) pulse train and (c) sampled output

The flat-top sampling technique is often used for voice signal sampling in a conventional PCM system. The implementation of flat-top sampling involves the use of a sample-and-hold circuit, as shown in Fig. 2.7(a). A sample-and-hold circuit converts a constantly changing analog signal into a sequence of PAM voltage levels with constant amplitude, as shown in Fig 2.7(b).

The field-effect transistor (FET) in the sample-and-hold circuit functions as an analog switch. When the FET is turned on, Q_1 will provide a low impedance path that will allow dumping across C_1, the analog sample voltage. The time over which Q_1 is turned on is known as aperture or acquisition time. Basically, C_1 in the Fig. 2.7(a) can be viewed as the hold circuit. On the other hand, if Q_1 is turned off, C_1 may not discharge because of the absence of a complete path and hence, will be storing the sampled voltage. The storage time relating to the capacitor is referred to as the analog-to-digital conversion time. This is because it is time during which the analog-to-digital convertor will convert the sample voltage into a PCM code.

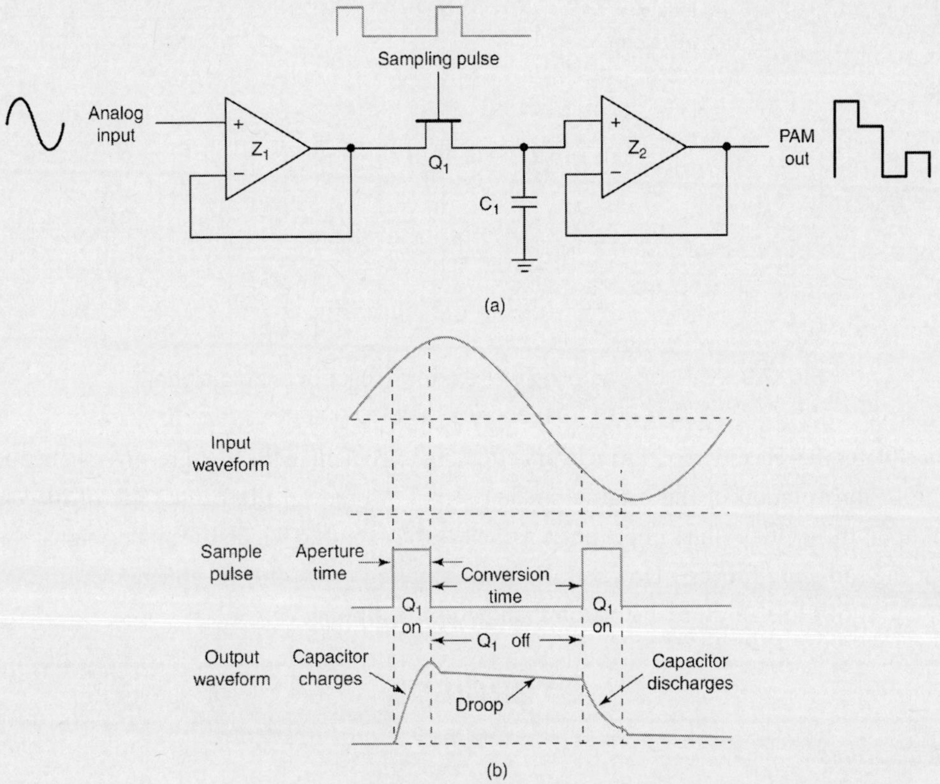

Fig. 2.7: (a) Sample- and- hold circuit and (b) waveforms exhibiting the sample- and- hold action

Fig 2.7(b) depicts the input signal, the sampling pulse and the waveform generated across the capacitor C_1. One important constraint is that the output impedance and the on resistance of the voltage follower Z_1 and Q_1, respectively, should be made as small as possible to ensure very short RC charging time constant of the capacitor. This will allow the capacitor to discharge or charge quickly given the short acquisition time. Furthermore, any swift drop in the voltage of the capacitor that follows each of the sample pluses is primarily because of the redistribution of charge across the capacitor C_1.

When the FET is off and if the capacitance between the FET's gate and drain is placed in series with C_1, then the circuit will act as a voltage-divider. It may also be noted that during the conversion time, there is a slow discharge across the capacitor due to the capacitor's own leakage resistance and the input impedance relating to the voltage follower Z_2. This phenomenon is called droop. Consequently, the leakage resistance of C_1 as well as the input impedance of Z_2 should be chosen to be as high as possible. Fundamentally, Z_1 and Z_2 are responsible for isolating Q_1 and C_1 from the input and output circuitry.

Flat-top sampling samples the input voltage with a short pulse and keeps it stable until the next sample. Fig. 2.8 illustrates the concept of flat-top sampling, a process in which the frequency spectrum is modified during sampling, resulting in the introduction of an error known as *aperture error*. This error happens when the amplitudes of the signal being sampled fluctuate throughout the sample pulse's duration.

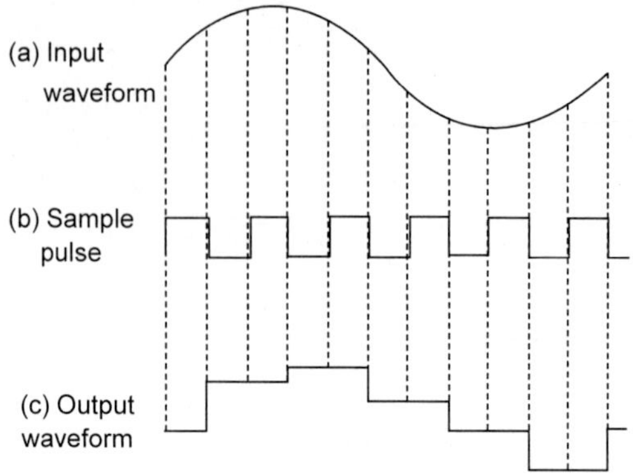

Fig. 2.8: Flat- top sampling waveforms: (a) input analog signal, (b) pulse train and (c) sampled output

This facilitates the relaxation of the recovery circuit requirement in the PCM receiver, hence enabling the accurate reconstruction of the original analog signal. Moreover, the magnitude of inaccuracy is contingent upon the width of the sample pulse as well as the extent to which the analog signal fluctuates throughout the sampling process. However, flat-top sampling offers the advantage of working with a slow ADC and shows less aperture distortion than natural sampling.

EXERCISES

Two mark questions

1. Define the sampling theorem.
2. Define the sampling rate.
3. What is the purpose of anti-aliasing filter?
4. How do you overcome the aliasing problem?
5. What is an aperture error?
6. Mention the differences between natural sampling and flat-top sampling.

Five mark questions

1. Explain the operation of a sample-and-hold circuit.
2. Describe the concept of interpolation.
3. Write a short note on pulse amplitude modulation.
4. Illustrate the natural and flat-top sampling process with help of waveforms.

Numerical Problems

1. A signal of 15 MHz is sampled at a rate of 30 MHz. What alias is generated?

Solution:

$$f_a = f_s - f_m = 30 \text{ MHz} - 15 \text{ MHz} = 15 \text{ MHz}$$

2. Find the Nyquist sampling rate of 20 sin(16π × 10^3 t).

Solution:

$$x(t) = 20\ \sin\left(16\pi \times 10^3 t\right)$$
$$= 20\ \sin\left(2\pi \times 8 \times 10^3 t\right)$$

Hence, $f_m = 8$ kHz

Sampling rate, $f_s = 2f_m = 16$ kHz.

KNOW MORE

- Harry Nyquist was born in Sweden in 1889. He received his B.S. and M.S. degrees in electrical engineering from the University of North Dakota in 1914 and 1915, respectively, and his Ph.D. degree in 1917 from Yale University. He worked for the American Telephone and Telegraph Company from 1917 to 1934, where conducted research on communications engineering in the context of image and voice transmission. He continued his research in the Bell Labs in systems and transmission engineering.
- To his credit, he was granted 138 U.S. patents and his main research interests were in thermal noise characterization and signal transmissions. He was the one who postulated the Nyquist sampling theorem which says that the sampling rate at which the analog signal is to be sampled for successful reconstruction at the receiver is that it should be at least twice the highest frequency component of the signal under consideration.
- He also formulated the mathematical model of thermal noise. These works indeed laid strong foundations for today's data transmissions and information theory. Additionally, his invention related to vestigial sideband transmission is being used widely in television transmission and the notable Nyquist plot is used for stability analysis of feedback systems. Furthermore, he has received numerous awards and medals to recognize for his contributions to fundamental engineering.
- The most widely employed sampling rates for audio are 44.1 kHz and 48 kHz. To be specific, 44.1 kHz is used in compact discs and 48 kHz in the production of video and film.
- For music, higher sampling rates, such as 88.2 kHz, 96 kHz and 192 kHz, are used for higher resolution.
- Interpolation is used in weather forecasts such as predicting rainfall.
- In image processing, quantization is employed to reduce the number of colors needed to describe an image in digital form.
- In digital audio, the aliasing effect is temporal while in digital images, it is spatial.

REFERENCES AND SUGGESTED READINGS

Taub and Schilling: *Principles of Communication Systems*, Tata McGraw-Hill, New Delhi, 1995.

Haykin S: Digital Communications, Wiley India, 2017.

References for Further Reading

1. https://archive.nptel.ac.in/courses/108/104/108104091/

2. http://vlabs.iitkgp.ernet.in/dsp/exp1/Exp1S/Exp1_S.html

3 PCM and Delta Modulation Systems

UNIT SPECIFICS

Through this unit, we shall discuss the following aspects:

- Benefits of PCM
- Process of quantization
- Uniform and non-uniform quantization: the *μ-law* and *A-law*
- Quantization noise
- Transmitter and receiver operations of PCM
- Delta modulator transmitter and receiver circuits
- Noises of delta modulation
- Analysis of signal-to-quantization noise ratio

RATIONALE

The unit begins with the discussion of benefits of PCM technique. The quantization process and its types are elaborated in detail with necessary formulae. The quantization noise and bit rate of PCM are also derived. The transmitter and receiver modules of PCM are explained in detail. Later, theory of transmission and reception of DM are discussed with neat block diagrams. Finally, signal-to-quantization ratio is analysed with examples.

PRE-REQUISITES

Knowledge of the sampling process

UNIT OUTCOMES

List of outcomes of this unit is as follows:

U3-O1: Understand the fundamentals of the quantization process.

U3-O2: Explain of uniform and non-uniform quantization types with necessary mathematical explanations.

U3-O3: Discuss the transmitter and receiver modules of PCM.

U3-O4: Study the transmitter and receiver block diagram of DM. Illustrate the noises of DM and suggest how to minimize them.

U3-O5: Analyse the signal-to-quantization noise ratio with illustrations.

Unit-3 Outcome	EXPECTED MAPPING WITH COURSE OUTCOMES (1 - Weak Correlation; 2 - Medium Correlation; 3 - Strong Correlation)				
	CO-1	CO-2	CO-3	CO-4	CO-5
U3-O1	1	1	3	1	-
U3-O2	1	1	3	1	-
U3-O3	1	1	3	1	-
U3-O4	1	1	3	1	-
U3-O5	1	1	3	1	-

MODULE 3: PCM AND DELTA MODULATION SYSTEMS

3.1 INTRODUCTION

This unit focuses on digital pulse modulation, which can be viewed as converting *analog signals into coded pulses*. To be specific, the conversion can be described as translation from analog to digital communications. The below modulation techniques describes the conversion process.

- *Pulse-code modulation* (PCM), as mentioned earlier, is a robust scheme, but demands in bandwidth that is required for transmission as well as in terms of computational efforts. However, PCM has been a widely accepted standard method for converting the voice and video signals into digital format. It is adopted in the telephone communication.
- *Delta modulation* (DM) is relatively a simple method and easy to realize but demands a considerable increase in bandwidth required for transmission.

Despite their inherent distinctions, PCM and DM share a significant common attribute: the message signal is characterised by discrete-time and discrete-amplitude. Pulse amplitude modulation (PAM) is responsible for the discrete-time characterization, while the representation of discrete-amplitude can be achieved through a process known as quantization, which will be discussed in the subsequent discourse. As mentioned earlier, this unit will examine two distinct forms of pulse modulation:

- PCM is widely regarded as a very appealing method for digitally transmitting analog signals carrying information, such as audio and video. This approach offers many notable benefits.
- *Robustness* in the context of interference and channel noise.
- The transmission connection enables the convenient and effective *regeneration* of the coded pulses.
- Results in increased channel bandwidth exchange that thereby resulting in enhanced signal-to quantization noise ratio, and comply with the exponential law.

- Supports *uniform configuration* for transmitting diverse types of baseband signals, and, hence its incorporation into a common network with other forms of digital data.
- Comparatively easier to drop or insert a message source in a multiplexing system.
- Provide *secure* communication when special modulation schemes or encryption is used.

DM addresses practical limitations of PCM such as increased system complexity and increased transmission bandwidth. When simplicity of implementation is the requirement, DM can be used which intentionally 'oversamples' the message signal. In the present scenario, the trade-off between enhancing transmission bandwidth and diminishing the complexity of the system is seen. Although the two analog-to-digital translation approaches vary, they both use two basic signal-processing procedures: sampling and quantization.

In order to provide more clarity, the first step involves conducting a sampling procedure, which is then followed by the application of PAM, and ultimately concludes with the process of quantization pertaining to amplitude. The sampling process, which is often explained in the temporal domain, is essential to both digital communications and digital signal processing. An analog signal is converted into a set of uniformly spaced samples during the sampling process. The bandwidth of the message signal must be carefully considered while choosing the sample rate to ensure practical efficacy. The main premise of the sampling theorem is that the original analog message signal may be uniquely characterized by the matching sequence of samples.

3.2 QUANTIZATION

An analog message signal consists of a continuous range of amplitudes, and hence, the samples of message signal constitute continuous amplitude range. When looking over a large range of signal amplitudes, an unlimited number of amplitude levels may be seen. However, since our hearing and visual sense can only detect limited variations in intensity, it is not required to relay all of the amplitudes of the samples. As a result, we may approximate the message signal by a signal that is constructed with the aid of discrete amplitudes from the available set with minimum error constraints.

In the context of waveform coding, such as PCM, a fundamental need is the presence of a limited set of discrete amplitude levels. By carefully selecting these discrete amplitude levels with small intervals between them, it becomes difficult to distinguish the approximated signal from the original signal. Hence, amplitude quantization, often known as quantization, may be described in the following manner.

The process of converting a message signal's sample amplitude, $m(nT_s)$, which is represented as $m(t)$ at time $t = nT_s$, into discrete amplitudes, $v(nT_s)$, chosen from a constrained range of possible amplitudes is known as quantization. The aforementioned definition assumes that the quantizer, which carries out the quantization procedure, is *instantaneous and memoryless*. Samples of the message signal $m(t)$ that come before or after have no effect on the conversion at time $t = nT_s$. Though not ideal, the proposed approach often uses a simple scalar quantization technique.

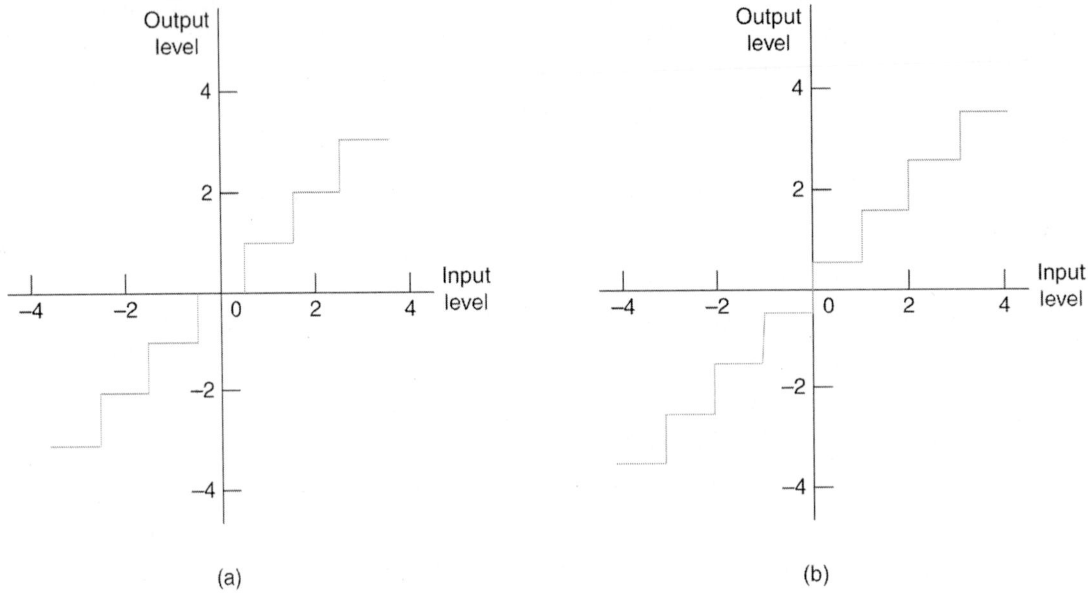

Fig. 3.1: Types of uniform quantization: (a) mid- tread (b) mid-rise

Quantizer is usually of two types namely *uniform or non-uniform*. The representation levels in a uniform quantizer are usually spaced uniformly. However, it is otherwise in the non-uniform quantizer. This section will examine the concepts of uniform quantizers and non-uniform quantizers. The quantizer feature may be classified into two types: *mid-tread* or *mid-rise type*. This uniform quantizer of the mid-tread type is shown with its input–output characteristic in Fig. 3.1(a). The term 'mid-tread' is used to describe a specific characteristic of the staircase graph, where the origin is positioned at the midpoint of a tread. Moreover, Fig. 3.1(b) illustrates the input-output feature of a mid-rise type uniform quantizer, where the origin is positioned at the midpoint of the ascending section of the stair case graph. It is important to acknowledge that both forms of uniform quantizers exhibit symmetry around the origin, despite their distinct visual characteristics.

3.2.1 Quantization Noise

The use of quantization will result in the introduction of an error. The disparity between the quantized output sample v and the continuous input sample m may be used to describe this error. The inaccuracy that arises as a consequence is sometimes referred to as *quantization noise*. Fig. 3.2 illustrates the temporal evolution of quantization noise for a mid-tread type uniform quantizer.

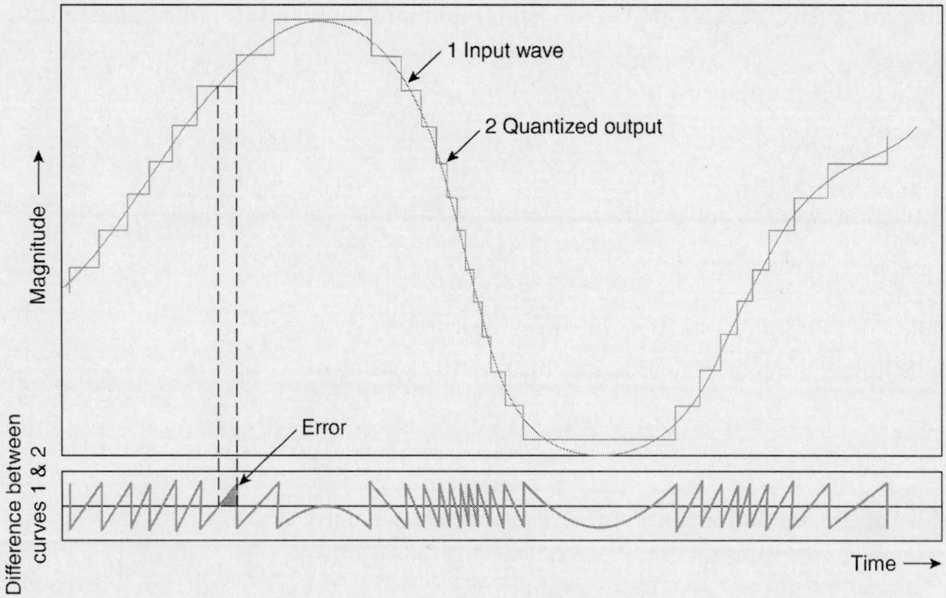

Fig. 3.2 Illustration of the quantization process with temporal evolution of quantization noise

If we represent the quantization error as the random variable Q with a sample value q, we may express $q = m - v$ or correspondingly to the given framework, the quantization error Q of a uniform quantizer will exhibit sample values that are constrained to the bounds of $-\Delta/2 \le q < \Delta/2$. Assuming a small step size (Δ), it can be inferred that the representation levels (L) will be correspondingly big.

Given this specific situation, it is plausible to assume that the quantization error Q is a random variable with a uniform distribution. Moreover, it is possible to draw comparisons between the effects of thermal noise and the quantization error on the quantizer's input. Therefore, the term *'quantization error'* is often used to denote the presence of *quantization noise.* If the quantization noise has a mean of zero, then the variance will be equal to the mean square value Q, which may be expressed as follows

$$\sigma_Q^2 = \frac{1}{\Delta}\int_{-\Delta/2}^{\Delta/2} q^2\, dq \tag{3.1}$$

$$= \frac{\Delta^2}{12}. \tag{3.2}$$

Now, let R represent the number of bits per sample that we may use to construct the binary code. Hence, we may write

$$L = 2^R \tag{3.3}$$

where L denotes the total representation levels. The above expression can also be written as:

$$\mathrm{L} = \log_2 L \tag{3.4}$$

Defining m_{max} and m_{min} as maximum and minimum amplitude of the input analog signal respectively, then,

$$\Delta = \frac{2m_{max}}{L}$$

alternatively, when both maximum and minimum amplitudes are same then,

$$\Delta = \frac{(m_{max} - m_{min})}{L} \tag{3.5}$$

and substituting the above equation into the step size results in

$$\Delta = \frac{2m_{max}}{2^R}. \tag{3.6}$$

Using the above step size in noise variance expression, we arrive at

$$\sigma_Q^2 = \frac{1}{3} m_{max}^2 2^{-2R} \tag{3.7}$$

Consider the following case, in which P is the message signal's average power and the output signal-to-noise (SNR) ratio of a uniform quantizer may be expressed as follows

$$(\text{SNR})_O \quad = \frac{P}{\sigma_Q^2} \tag{3.8}$$

$$= \left(\frac{3P}{m_{max}^2}\right) 2^{2R}. \tag{3.9}$$

The PCM bit rate is defined as product of the no. of samples and no. of bits required to represent a sample which is given by PCM bit rate, $BR = f_s R$.

The minimum bandwidth required to transmit a PCM signal is $B_T > BR/2$.

3.2.2 Non-uniform Quantization

The process of quantization may adhere to a consistent law, as previously discussed. In the context of telephonic communication, it is advantageous to include a flexible degree of differentiation across levels of representation in order to optimise the efficiency of the communication channel. To have a better understanding, let us consider the example of quantizing the voice signals that has the voltages range that spans from the peaks to weak passage (loud to weak talk) in the order of 1000 to 1. In this particular case, it is possible to use a non-uniform quantizer that incorporates a step size, which progressively enlarges as the spacing between the input–output amplitude characteristic and the origin of the quantizer expands.

The quantizer's large step size effectively handles any deviations of the speech signal into high amplitude bands, which often occur sporadically. In other terms, it might be posited that the vulnerable sections requiring heightened safeguarding are bolstered at the expense of the audibly prominent sections. By this method, we can achieve a nearly unvarying percentage exactness throughout the larger part of the input signal amplitude range. In addition, this results in the use of fewer steps that will be needed for the quantization process than a uniform quantizer. Consequently, we end up with better utilization of the channel. Furthermore, in the case when memoryless quantization is assumed, utilising a non-uniform quantizer may be compared to sending the message signal via a compressor and then to a uniform quantizer, as shown in Fig. 3.3.

One often used the compression law in practical applications is known as μ-*law*. This law is expressed as follows:

$$|v| = \frac{\ln(1+\mu|m|)}{\ln(1+\mu)}. \tag{3.10}$$

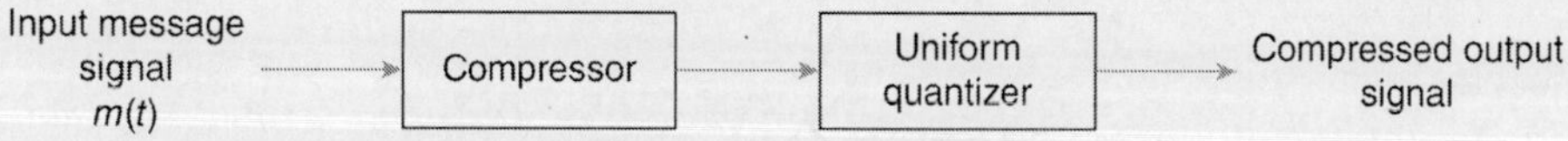

Fig. 3.3: Non- uniform quantization of the input message signal $m(t)$

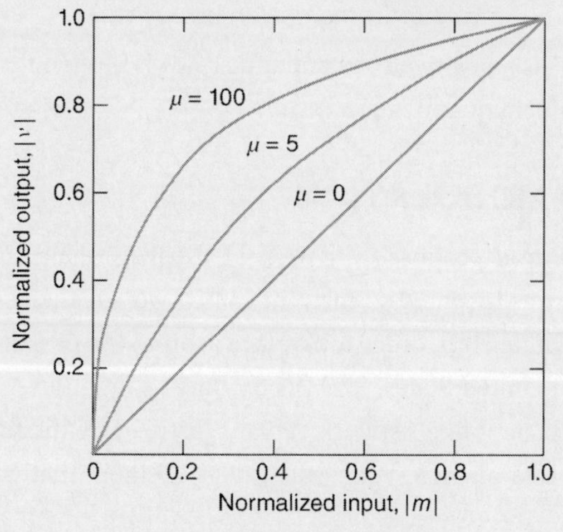

Fig. 3.4: Compression law: μ- law

In the aforementioned formula, the ln or $\log_e$, represents the natural logarithm. The compressor's input and output voltages are represented by the variables m and v, respectively, whereas μ is a positive constant. It should be noted that there is an assumption that the variables m and v be scaled, resulting in their values being restricted to the interval of $[-1, 1]$. Fig. 3.4 displays the μ-*law* for three distinct values of μ. Also, note that, $\mu = 0$ corresponds to the case of uniform quantization.

A-law is the frequently used compression algorithm, which is expressed as follows:

$$|v| = \begin{cases} \dfrac{A|m|}{1+\ln A}, & 0 =|m|= \dfrac{1}{A} \\ \\ \dfrac{1+\ln A|m|}{1+\ln A}, & \dfrac{1}{A} =|m|= 1 \end{cases} \tag{3.11}$$

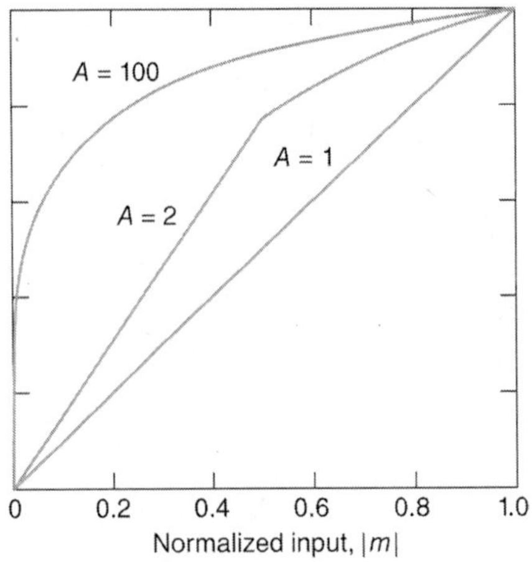

Fig. 3.5: Compression law: *A*- law

In the above expression, A is a positive constant and A-law is plotted for varying A in Fig. 3.5. $A = 1$ corresponds to the case of uniform quantization.

3.3 PULSE-CODED MODULATION

In 1937, the PCM technique was proposed by Alex H. Reeves when he was employed at AT&T's laboratory in Paris. Though it was invented so early, its merits were realized only in the mid-1960s. This is because the advent of solid-state electronics took place around the 60s, after which the PCM became widespread. Even today in the United States, PCM is the most sought-after communication method in the context of public switched telephone network. This is justified by the fact that PCM can effectively merge digital voice and data into a single, high-speed digital signal that can be sent across optical or metallic fibre lines.

Fig. 3.6 illustrates the PCM a prevalent method of digitally encoded modulation used in the realm of digital communication. It is important to acknowledge that the name 'pulse code modulation' might be deceptive, since it does not refer to a modulation technique, but rather to a method of digitally encoding analog signals. In the context of PCM, the duration and magnitude of the pulses are predetermined and unchangeable. Moreover, within the context of PCM, the existence or absence of a pulse inside a designated time interval signifies a logical value of 1 or 0, respectively.

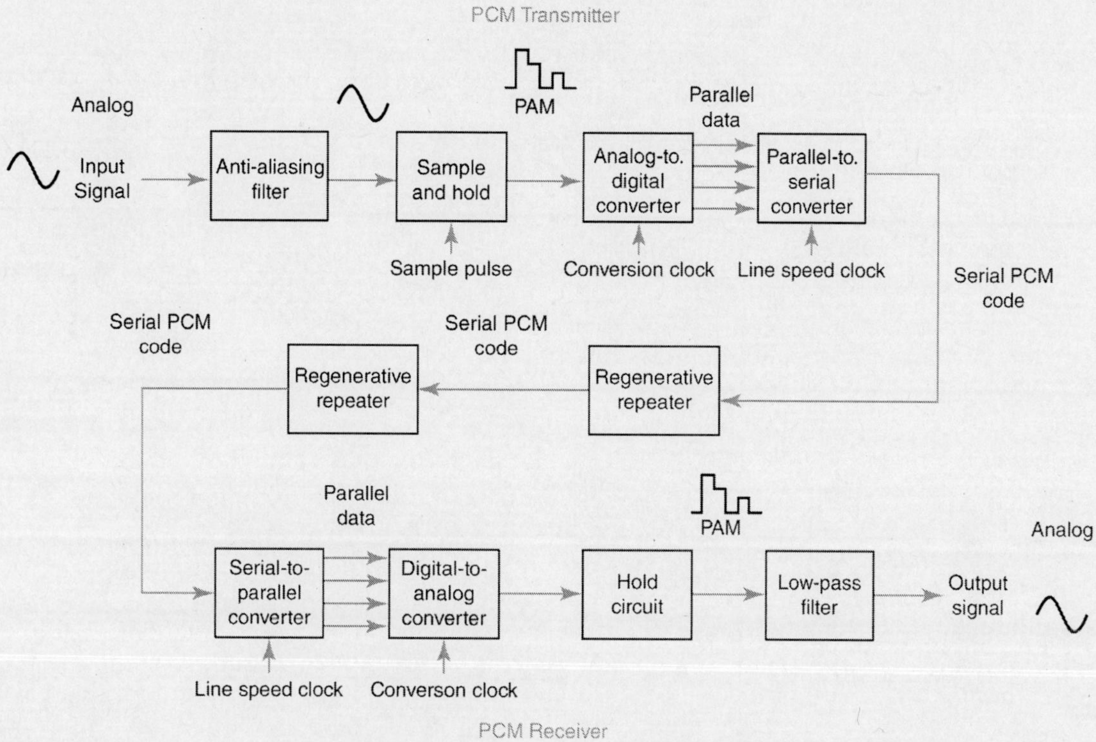

Fig. 3.6: PCM transmitter and receiver

The block diagram of a simplex communication system, namely a single-channel one-way PCM system, is shown in Fig. 3.6. The input signal undergoes band limitation by the implementation of a band pass filter, resulting in a frequency range that is often associated with human vocalization, specifically ranging from 300 Hz to 3000 Hz. The *sample-and-hold circuit* of the transmitter will periodically take samples of the analog input signal and transform them into a multi-level PAM signal. The PAM samples are converted into PCM codes by the analog-to-digital convertor (ADC).

This PCM code is then converted by the *parallel-to-serial converter* into a binary data stream. Next, a sequence of digital pulses is sent across the channel as the parallel-to-serial converter's output. The *repeaters* are deployed at a required distance on the channel to regenerate the digital pulses which otherwise would get attenuated when transmitted over long distances. At the recipient end, the serial-to-parallel converter performs the task of converting the incoming sequence of pulses into parallel PCM codes. Subsequently, the digital-to-analog convertor (DAC) is responsible for converting these parallel PCM codes into multi-level PAM signals. The hold circuit, seen in the block diagram, functions as a low-pass filter, facilitating the conversion of PAM signals into their respective analog counterparts. The output waveforms of PCM and other modulation techniques, such as PAM, pulse-width modulation (PWM) and pulse-position modulation (PPM), are shown in Fig. 3.7.

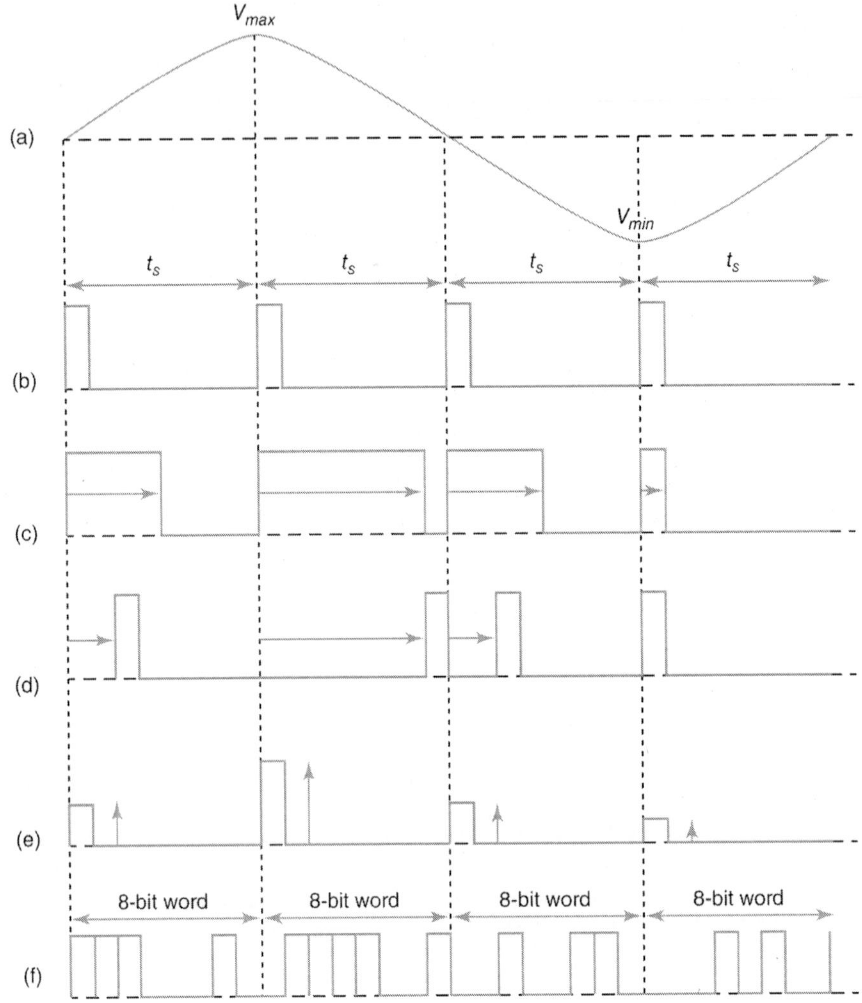

Fig. 3.7: Pulse modulation: (a) analog signal, (b) sample pulse, (c) PWM, (d) PPM, (e) PAM and (f) PCM

3.4 DELTA MODULATION

DM is a kind of PCM that employs a single-bit coding to facilitate the conversion of analog signals into digital format for transmission purposes. Since multi-bit codes must comprise a variety of values related to a sample, they are required in the context of standard PCM in order to appropriately represent both the sign and magnitude of a particular sample. As opposed to utilising a coded representation of the sample as seen in PCM, in the context of DM, the transmission of a single bit is sufficient to signal whether a given sample is more or smaller in magnitude than the preceding sample. The approach for a DM encoding scheme is characterized by a straightforward process. In this scheme, if the current sample is less in magnitude compared to the previous sample, a logic 0 is broadcast. On the other hand, a logic 1 is transmitted if the current sample has a greater magnitude than the preceding sample.

3.4.1 Delta Modulation Transmitter

A standard delta modulation transmitter is shown in the block schematic form in Fig. 3.8. The analog input signal must first be sampled and transformed into a PAM signal. The regenerated magnitude is contrasted to the DAC output voltage, which indicates the magnitude of the prior sample. The up-down counter stores this regenerated magnitude as a binary number. The up-down counter in a delta modulator is synchronized with the sample rate and has its value changed either upwards or downwards based on whether the sample before it was larger or smaller than the present sample.

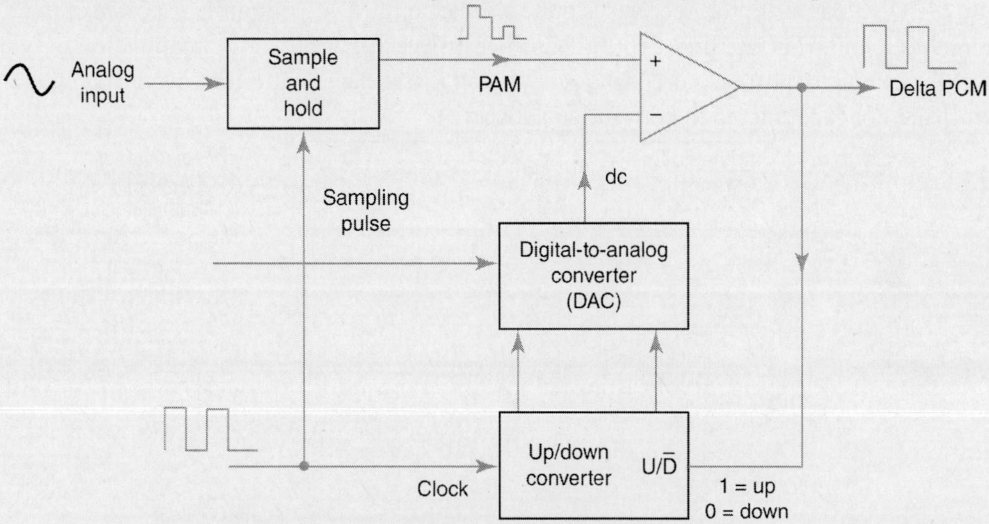

Fig. 3.8: DM transmitter

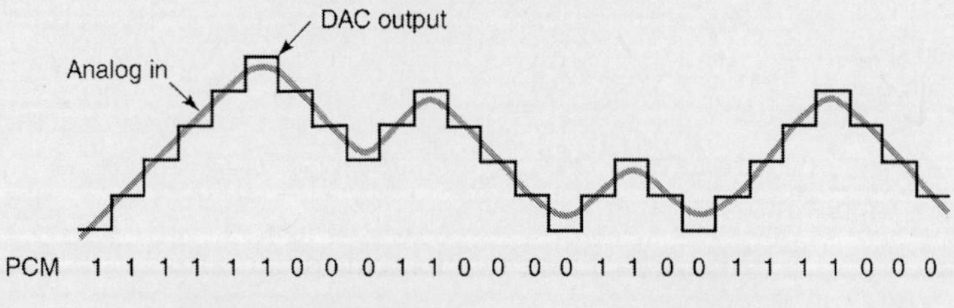

Fig. 3.9: Ideal operation of a DM encoder

Fig. 3.9 illustrates the ideal operation of a delta modulation encoder and illustrates the basic workings of a delta modulator encoder. When the up-down counter is first reset to zero, the DAC output is zero volts. The original sample is then evaluated and transformed into a PAM signal, which is then contrasted with a 0 V reference voltage. The current sample has a larger amplitude than the preceding sample when the comparator's output is logic 1 (+V). During the subsequent clock pulse, the up–down counter will undergo an increment operation, resulting in a count of 1. Consequently, the DAC will provide an output voltage that matches the minimal step size in terms of magnitude.

It is important to observe that within a delta modulator, the magnitude of the step size varies in proportion to the sample rate. As a result, until the DAC's output exceeds the amplitude of the analog sample, the up-down counter will follow the analog input signal. The up–down counter will then start counting down until the DAC output is less than the sample's amplitude. It is anticipated that the DAC will accurately reproduce the analog input signal in the hypothetical scenario shown in Fig. 3.9. More precisely, a logic 1 is generated when the up-down counter rises, and a logic 0 is conveyed when the counter decreases.

3.4.2 Delta Modulation Receiver

A standard receiver block design of delta modulation is shown in Fig. 3.10. The receiver has a high degree of resemblance to the transmitter, with the notable exception of the inclusion of a comparator. The up–down counter increments or decrements are based on whether logic 1 or logic 0 is received. Therefore, the DAC output in the receiver is quite similar to the DAC output in the transmitter. Delta modulation has a lower bit rate than typical PCM because each sample in the modulation only needs one bit to be transmitted. Furthermore, delta modulation is associated with slope overload and granular noise, which are not encountered in conventional PCM.

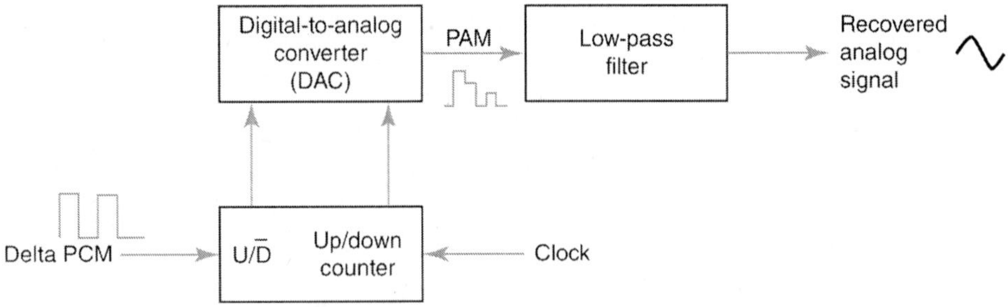

Fig. 3.10: DM receiver

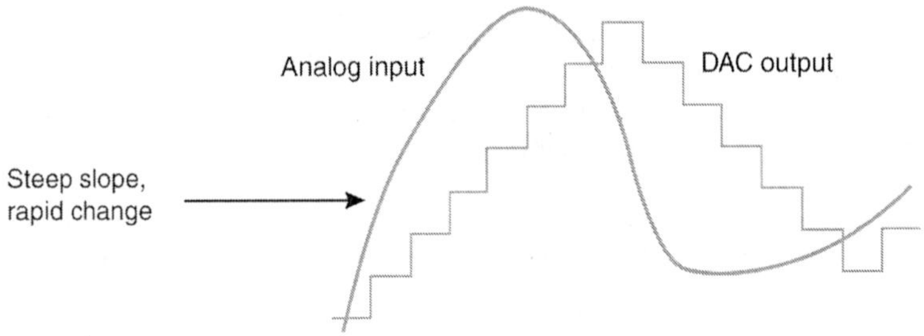

Fig. 3.11: Slope overload distortion

3.4.3 Slope Overload Noise

This effect of slope overload noise is illustrated in Fig. 3.11. In this context, the consequences of a rapid shift in the input signal exceeding the capabilities of the DAC are evident. In the current case, the input signal's magnitude of slope is greater than the delta modulator's tracking accuracy, leading to a phenomenon often known as slope overload. Increasing the minimum step size degree and clock frequency are two possible ways to mitigate slope overload.

3.4.4 Granular Noise

The granular noise is depicted in Fig. 3.12. The picture illustrates that the analog input signal exhibits consistent amplitude. However, the reconstructed signal displays fluctuations that are not inherent to the original input signal. The phenomenon in question is often referred to as granular noise. The granular noise seen in DM has similarities to the quantization noise often encountered in PCM systems. The reduction of the step size may effectively minimize the granular noise. Thus, a lower resolution reduces granular noise while a higher resolution reduces slope overload noise. Granular noise is common in analog signals with gentle slopes and amplitude changes. In contrast, slope overload is a common occurrence in analog signals characterized by sharp slopes or quick changes in amplitude.

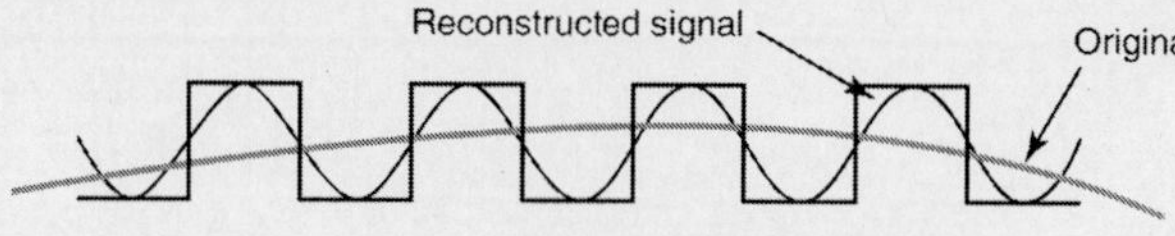

Fig. 3.12: Granular noise

3.5 SIGNAL-TO-QUANTIZATION NOISE RATIO

Figs. 3.13 and 3.14 show a three-bit PCM coding scheme. The present context encompasses linear codes. This implies that the difference in magnitude between any two successive codes remains constant. As a result, the maximum value of the quantization noise is equal to half the resolution, or quantum value, and their quantization error in terms of magnitude also maintains the same. Consequently, the *signal voltage-to-quantization noise voltage ratio* (*SQR*) in the worst possible case occurs if the input signals have minimum amplitude, i.e., 101 or 001. In this context, *SQR* worst-case is given by

$$SQR = \frac{resolution}{Q_e} = \frac{V_{lsb}}{V_{lsb}/2} = 2. \quad (3.12)$$

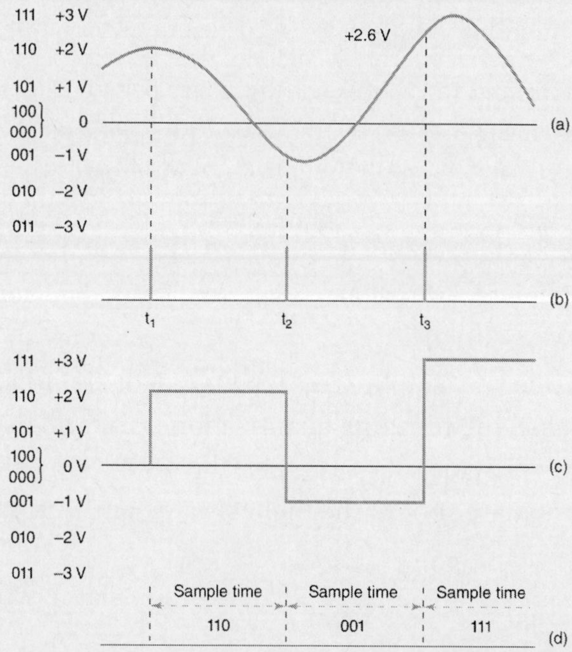

Fig. 3.13: (a) analog input signal; (b) sample pulse; (c) PAM signal; and (d) PCM code

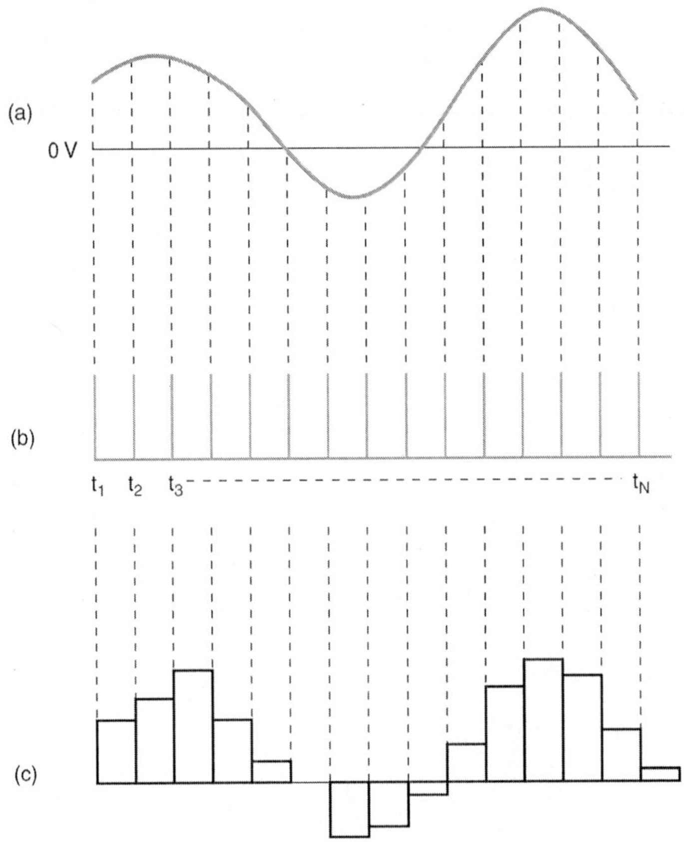

Fig. 3.14: PAM waveforms: (a) input signal, (b) sample pulse and (c) the PAM signal

The PCM code exampled in Fig. 3.13 displays the worst-case *SQR* for the code presented in Fig. 3.13, when the quantization voltage magnitude is the lowest, i.e., ±1 V, and hence the minimum *SQR* is

$$SQR_{(min)} = \frac{1}{0.5} = 2 \tag{3.13}$$

$$= 20\log(2)$$

$$SQR_{(min)}\text{dB} = 6\text{ dB}. \tag{3.14}$$

Suppose, if we consider an amplitude of 3 V (111 or 011) i.e., If the input signal reaches its maximum value, the resulting quantization noise will be equivalent to half of the resolution. In order to provide precise details, the resolution will be halved, resulting in the calculation of the square root of the highest value of the input signal.

$$SQR_{(max)} = \frac{V_{max}}{Q_e} = \frac{3}{0.5/2} = 6 \tag{3.15}$$

$$= 20 log\, 6.$$

$$SQR_{(max)}\text{dB} = 15.6\,\text{dB}. \tag{3.16}$$

The example above demonstrates that although the quantization error size is constant throughout the PCM coding, the error percentage fluctuates and becomes less as sample size increases. Furthermore, the earlier *SQR* formulation assumed the maximum quantization error and related to voltage. It is thus mostly provided for comparison and has little use in real life. It also helps in proving that *SQR* is not a stationary object across the complete amplitude range of the sample.

However, within the context of realism and as shown in Fig. 3.14, it can be seen that the PAM waveform and the input waveform exhibit dissimilarities due to variations in their magnitudes. As a result, the *SQR* is not steady. The quantization error from digitalizing an analog signal sample is generally measured by comparing the average signal power to the noise power. Furthermore, within the context of linear PCM, it is seen that all quantization intervals possess identical magnitudes. As such, the following mathematical expression determines *the SNR ratio also known as the signal-to-distortion ratio* or *signal -to-quantizing noise ratio* to be evaluated:

$$SQR_{(\mathrm{dB})} = 10\log\frac{v^2 / R}{\left(q^2/12\right)/R} \tag{3.17}$$

where

R = resistance (Ω)

v = rms signal voltage (V)

q = quantization interval (V)

v^2/R = average signal power (W)

$(q^2/12)/R$ = average quantization noise power (W)

Suppose, if we assume the resistance values are equal, the above expression will reduce to

$$SQR_{(\mathrm{dB})} = 10\log\left[\frac{v^2}{\frac{q^2}{12}}\right] \tag{3.18}$$

$$= 10.8 + \left[20\log\frac{v}{q}\right]. \tag{3.19}$$

EXERCISES

Two mark questions

1. Define quantization.
2. Write the differences between mid-tread and mid-rise quantizers.
3. Mention a few applications of the *μ-law and A-law*.
4. When will you apply non-uniform quantization?
5. What are modules of a PCM transmitter?

6. How will you derive quantization noise power?
7. How will you minimize the slope overload noise?
8. Why do we call DM as a single-bit quantizer?
9. Which are the two types of noises associated with DM?
10. Mention the practical applications of PCM.

Five mark questions

1. Elaborate on the signal-to-quantization noise ratio.
2. Bring out the distinction between uniform and non-uniform quantization.
3. Discuss the uniform and non-uniform quantizers with necessary equations.
4. Brief the discussion on the signal-to-quantization noise ratio of PCM.

Ten mark questions

1. Discuss in detail various modules of pulse-coded modulation transmitter and receiver with neat diagrams.
2. Explain the transmitter and receiver block diagram of delta modulator in detail. Also elaborate two different types of noises associated with it and also mention how to overcome those noises.

Numerical Problems

1. An analog signal of bandwidth 4 kHz is to be converted to the digital signal. If we use PCM and no. of bits required to represent a sample is 8 bits, What would be the bit rate?

Solution:

$f_m = 4$ kHz, sampling frequency $f_s = 2f_m = 8$ kHz.

PCM bit rate $= f_s{\cdot}R = 8$ kHz $\times\ 8 = 64$ kbps.

2. The bandwidth of the speech signal is 8 kHz and dynamic range of the signal is +5 and −5 V is sampled at the Nyquist rate. Each sample is quantized and encoded by 8 bits. How much is the SNR of the output PCM signal?

Solution:

$f_m = 8$ kHz, $m_{max} = 5$ V, $R = 8$ bits, $P = m^2{}_{max} = 25$ watts

$$(SNR)_O = \left(\frac{3P}{m^2_{max}}\right)2^{2R}$$

$$= \frac{3\times 25}{25}2^{2\times 8}$$

$$= 1{,}96{,}608$$

$$= 10\log(1{,}96{,}608) = 52.93 \text{ dB}.$$

KNOW MORE

Alec Harley Reeves, a British scientist, was born in Redhill, Surrey, in 1902 and holds 82 patents. He was renowned for his invention of pulse-code modulation (PCM). He graduated from Reigate Grammar School, City and Guilds Engineering College in 1918 and with a postgraduate from Imperial College London in 1921. While working for Western Electric Company in 1923, he along with his team was responsible for deploying the first transatlantic telephone link. Following this, in 1927, moved to Paris as ITT acquired Western Electric's European arm. In Paris, at the ITT lab, he worked on numerous projects such as the deployment of a short-wave radio link that connected the telephone networks between Spain and South America, the world's first-ever SSB radiotelephone transmission, multi-channel carrier development for radio telephones that operated in the UHF band. He also contributed to the innovations in digital delay links, and microphones based on condenser and circuit design for automatic frequency control. Alec Harley Reeves was celebrated as the 'Father of the Information Age', and was honoured in 1965 with the Stuart Ballantine Medal.

- PCM is employed as a coding technique in CDs, Blu-ray discs, and Red Book CDs.
- NHK's research lab in Japan developed the first PCM recorder.
- A facsimile machine that transmitted signals using a 5-bit PCM encoded by an opto-mechanical ADC was patented by Paul M. Rainey in 1926.
- During World War II, the encryption device named SIGSALY transmitted the first-ever high-level communications of speech using digital techniques.
- A working PCM system was built by Ferranti Canada that transmitted over long distances the digitized radar data.
- In 1952, F. de Jager of Philips Research Laboratories coupled oversampling and feedback to realize delta modulation.
- Satellite Business Systems used DM to provide long-distance telephone service over large domestic business corporations like IBM that were in need of a significant inter-corporation information exchange.

REFERENCES AND SUGGESTED READINGS

1. Wayne Tomasi: *Electronic Communication System*, Pearson Education, 5th edition, 2008.
2. Taub and Schilling: *Principles of Communication Systems*, Tata McGraw-Hill, New Delhi, 1995.
3. Simon Haykin: *Communication Systems*, Wiley Eastern, 4th edition, 2001.

References for further reading

1. https://archive.nptel.ac.in/courses/108/104/108104091/

2. https://archive.nptel.ac.in/courses/108/104/108104091/

3. https://www.youtube.com/watch?v=t6s5MTzYOWs

❑❑

4 Digital Modulation

UNIT SPECIFICS

Through this unit, we will discuss the following aspects:

- Fundamentals of line coding techniques
- Significance of pulse shaping and its techniques
- Signal design for achieving zero inter-symbol interference (ISI)
- Ideal Nyquist pulse for distortionless baseband data transmission
- Geometric interpretation of signals
- Gram–Schmidt orthogonalization procedure.

RATIONALE

This unit on begins with line coding which is a critical aspect of digital communication that involves mapping digital data into a digital signal. The basis for understanding the fundamentals of line coding lies in its role in mitigating errors during data transmission. Next, pulse shaping is crucial for controlling the spectral characteristics of a signal and ensuring efficient use of bandwidth. Various pulse-shaping techniques, such as raised cosine and Gaussian filters, allow engineers to tailor the signal to meet specific communication system requirements, ensuring reliable and efficient data transmission. Then, inter symbol interference (ISI) can degrade the performance of a communication system by causing overlap between adjacent symbols. The rationale for signal design aimed at achieving zero ISI is to ensure accurate symbol detection at the receiver.

Ideal Nyquist pulse is essential for achieving maximum data transmission rates while avoiding ISI. The Nyquist pulse ensures efficient use of bandwidth and aids in recovering the original data at the receiver with minimal distortion. The geometric interpretation of signals provides a visual understanding of signal characteristics in a communication system. This approach aids in comprehending concepts such as modulation, signal space, and constellation diagrams. A thorough understanding of these key concepts in digital communication is essential for designing and implementing robust communication systems.

PRE-REQUISITES

Class 10 Physics and Mathematics

UNIT OUTCOMES

List of outcomes of this unit is as follows:

U4-O1: Understand the fundamentals of line coding techniques.

U4-O2: Explain concept of pulse shaping and its techniques.

U4-O3: Study the Signal design for achieving zero-ISI.

U4-O4: learn the ideal Nyquist pulse for distortionless baseband data transmission.

U4-O5: Analyse the geometric interpretation of signals and Gram–Schmidt orthogonalization procedure

Unit-4 Outcome	EXPECTED MAPPING WITH COURSE OUTCOMES (1 - Weak Correlation; 2 - Medium Correlation; 3 - Strong Correlation)					
	CO-1	CO-2	CO-3	CO-4	CO-5	CO-6
U4-O1	1	-	1	3	-	1
U4-O2	1	-	1	3	-	1
U4-O3	1	-	1	3	-	1
U4-O4	1	-	1	3	-	1
U4-O5	1	-	1	3	-	1

MODULE 4: DIGITAL MODULATION

4.1 LINE CODES

Pulse code modulation (PCM) and delta modulation (DM) are the two coding schemes discussed in the previous chapter. Though these schemes differ naturally in several requirements such as transmission–bandwidth, the configuration of the transmitter–receiver, quantization noise, and complexity, both waveform-coding methods need *line codes* to represent the binary streams that are encoded by the respective transmitters. This will facilitate the binary streams to be transmitted over the communication channel. Fig. 4.1, below illustrates the waveforms corresponding to the five principle line coding techniques for the considered binary data, i.e., 01101001. Also, for positive frequencies, in Fig. 4.2, the respective power spectra that correspond to the binary stream that is generated randomly with the assumptions that the symbol 0 and symbol 1 are equiprobable, normalized average power to be unity, with the frequency f that is normalized relating to the bit rate $1/T_b$ are shown. In the forthcoming section, we explain the five line coding schemes that are involved in the coded waveforms generation.

4.1.1 Unipolar NRZ Signalling

In this type of line coding, symbol 1 and symbol 0 are represented by transmitting a pulse with amplitude A and switching off the pulse, respectively, for the pulse duration as shown in Fig. 4.2(a). Furthermore, this type of signalling scheme is also called *on–off signalling*. Though this scheme appears simple, it has two disadvantages namely wastage of power because of the DC that is transmitted and non-zero power spectrum at zero frequency with respect to the transmitted signal.

4.1.1.1 Polar NRZ Signalling

As illustrated in Fig. 4.2(b), in this signalling scheme, symbol 1 and symbol 0 are described by transmitting a pulse with amplitude $+A$ and a pulse with amplitude $-A$, respectively, and this line code is comparatively easier to generate. The main drawback of this line code is that it results in a non-zero power spectrum relating to the signal that is large enough close to zero frequency.

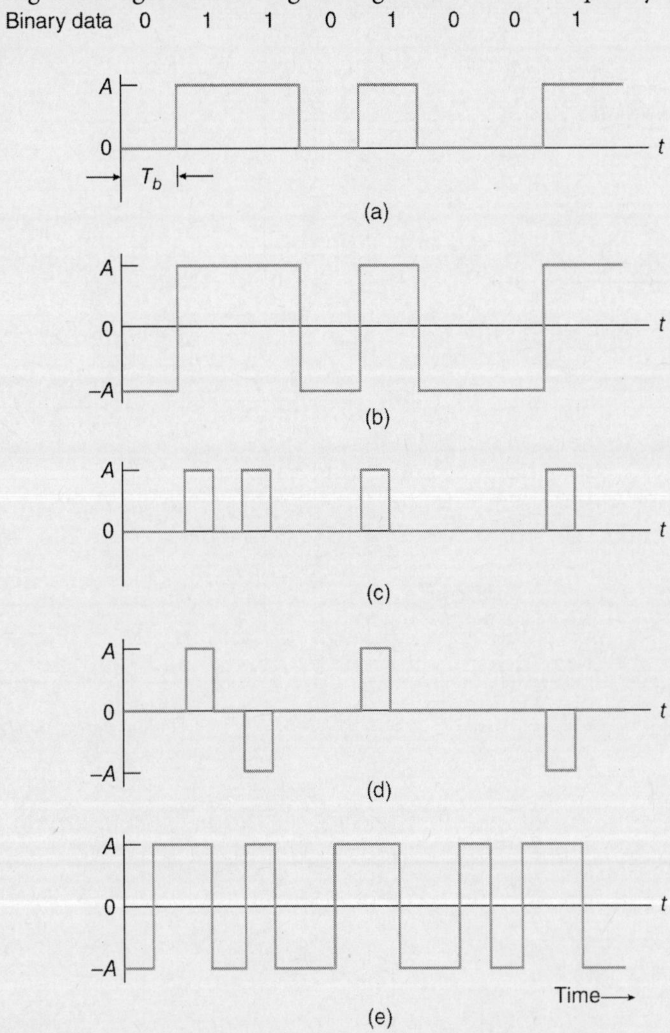

Fig. 4.1: Representations of the five principle lines codes electrically: (a) non- return- to- zero (NRZ) unipolar code, (b) NRZ polar code, (c) return- to- zero (RZ) unipolar code, (d) RZ bipolar code and (e) split-phase or Manchester code.

4.1.2 Unipolar RZ Signalling

In this type of signalling, symbol 1 is characterized by transmitting a rectangular pulse that is half symbol width and amplitude A while symbol 0 with the transmission of *no* pulse, as depicted in Fig. 4.5(c). Moreover, in this line coding scheme, in the spectrum relating to power of the signal that is transmitted, the existence of delta functions at frequencies $f = 0, \ \pm 1/T_b$ serves as an appealing feature that can be utilized at the receiver to perform *bit-timing* recovery. Nevertheless, the disadvantage of this scheme is that it needs 3 dB excess power compared to the polar RZ scheme to achieve similar symbol error probability.

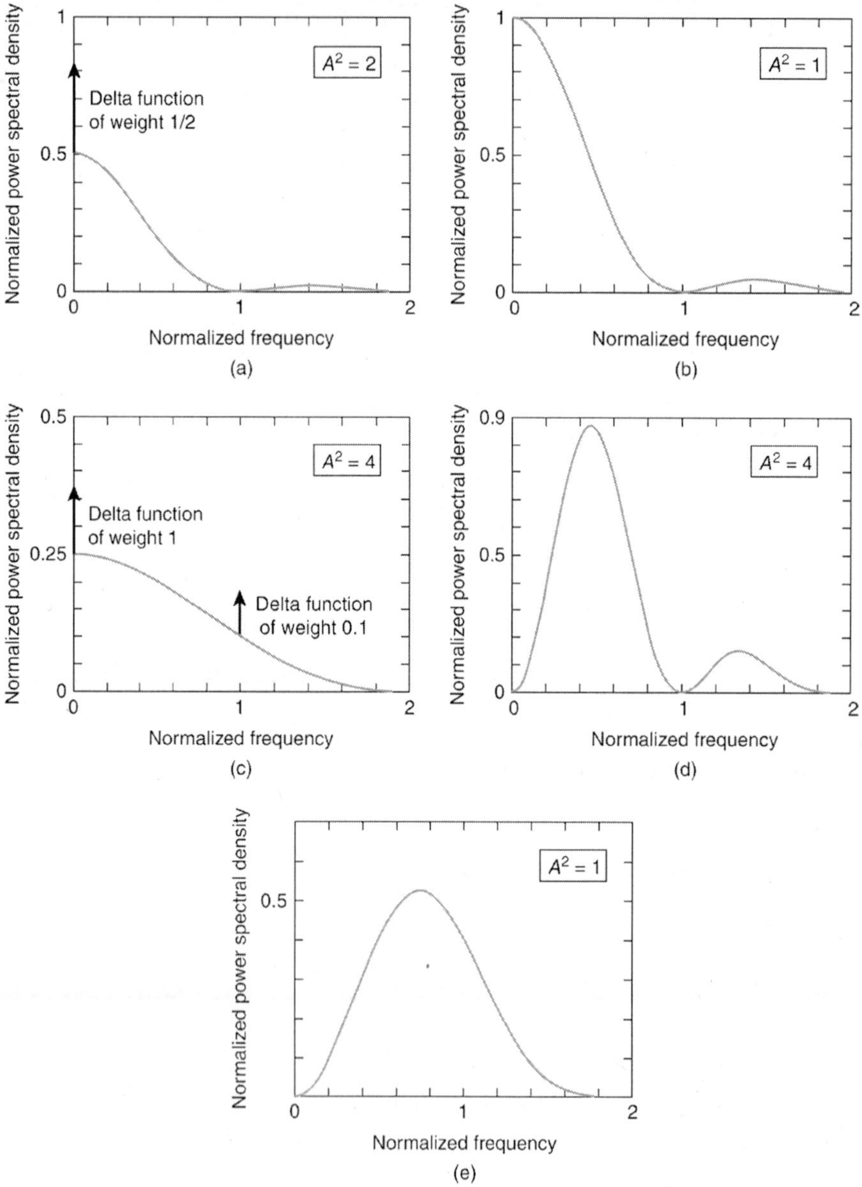

Fig. 4.2: Power spectra of line codes: (a) NRZ signal – unipolar, (b) NRZ signal – polar, (c) RZ signal – unipolar, (d) RZ signal – bipolar and (e) Manchester- encoded signal. In the plots, the frequency and the average power are normalized relating to the bit rate $1/T_b$, and unity, respectively.

4.1.2.1 Bipolar RZ Signalling

As shown in Fig. 4.5(d), this line coding scheme uses three levels of amplitude. To be specific, an equal amplitude positive pulse ($+A$) and negative pulse ($-A$) with each of the pulses having a duration equal to half symbol width are alternately used to represent symbol 1, and no pulse to represent symbol 0. This line coding scheme results in a power spectrum related to the signal that is transmitted, which indeed has relatively no DC component. Additionally, low-frequency components are inconsequential if the symbols 1 and 0 are equiprobable. Moreover, this line coding scheme is also known as alternate mark inversion (AMI) coding.

4.1.2.2 Split-phase or the Manchester Code

In this type of coding as evidenced in Figure, a pulse with positive amplitude A and a pulse with negative amplitude ($-A$), with each of the pulses having a duration equal to half symbol width is used to represent symbol 1, and reversed polarities of these two pulses are used to represent symbol 0. A distinctive quality of this split-phase coding is that the DC component is suppressed to a great extent. Additionally, this coding results in low-frequency components that are relatively insignificant, irrespective of the statistics of the transmitted signal, and this feature is crucial in some of the applications.

4.2 PULSE SHAPING

In the context of information communications, *'pulse shaping'* refers to modifying a pulse waveform of the transmitted signal in order to optimize it for transmission over a communication channel. This is usually carried out by limiting the bandwidth required for transmission followed by the process of filtering the pulses so that inter-symbol interference (ISI) is minimized. Pulse shaping is employed following the line coding and modulation and is important especially in RF communication so as to limit the signal within the frequency band under consideration.

4.2.1 Need for Pulse Shaping

If a higher modulation rate signal is transmitted over a channel that is band-limited, it can lead to ISI. This can be well understood from the Fourier transform wherein, a signal that is band-limited corresponds to an infinite time series resulting in the smearing of adjacent pulses. Furthermore, a higher modulation rate requires increased bandwidth. If we consider an ideal rectangular spectrum, its time domain equivalent is a sinc pulse. Suppose, the signal bandwidth is larger than the bandwidth of the channel, it may lead to distortions in the received signal, and this distortion is usually manifested as ISI. Theoretically speaking, if we use a sinc pulse, and if the adjaçent pulses are aligned perfectly at the zero-crossings, there will be no ISI. Nevertheless, to achieve this, perfect sampling and synchronization with no jitters are required. In practice, the eye diagram and its pattern are analysed to find out the presence of ISI. The eye pattern also can be used to visualize the effects due to channel, frequency, or synchronization stability.

The spectrum of the signal is commonly determined based on the modulation format utilized and the data rate requirement. However, we can modify it by applying pulse shaping which in turn will lead to a signal that is time-limited again by making the spectrum smooth. In fact, the symbols that are transmitted are usually represented as time-sequenced Dirac delta pulses that are multiplied by means of the symbol. This process can be viewed as a transition in the conventional sense from digital-to-analog and at this juncture, signal has unlimited bandwidth. This unlimited bandwidth signal is then filtered by invoking a pulse-shaping filter to produce the signal that was transmitted. In the time domain, suppose, if this pulse shaping filter is rectangular, then this may result in a spectrum that is unlimited.

In most of the typical communication systems, at the baseband, implicitly a boxcar filter is the pulse-shaping filter. When Fourier transform is applied, it takes the shape of $\sin(x)/x$. This indeed has considerable power pertaining to the signal at frequencies that are high enough than the symbol rate and usually does not present as a major predicament if twisted-pair or optical fibre is exploited as the channel. On the other hand, in the context of RF communications, which impose strict bandwidth restrictions for single transmissions, this will turn out into an issue that would result in wastage of bandwidth. To put it differently, we can say that the channel is band-limited. Consequently, we need to develop better filters that will minimize the bandwidth requirement for a specific symbol rate.

As another example, the pulse generation in electronics that would require short rise-time. One way to achieve this is to initiate with a slow-rising pulse, and reduce the rise-time with the help of a diode step-recovery circuit.

4.2.2 Pulse-shaping Filter

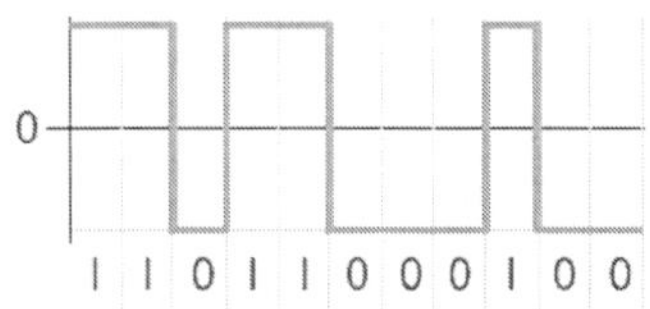

Figure illustrates a typical NRZ waveform-coded signal which is filtered with a sinc filter implicitly. Note that, every filter cannot be used as a pulse-shaping filter. The basic requirement is that the filter by and of itself should not introduce ISI, i.e., it should satisfy certain criteria. The most commonly adopted criterion is the Nyquist ISI as it correlates the transmitter signal's spectrum with the ISI. The following are the most commonly employed pulse-shaping filters in a typical communication system.

- Sinc-shaped filter
- Raised-cosine filter
- Gaussian filter

Usually, sending side pulse shaping is blended with a receiver side-matched filter to attain the most favourable tolerance for noise in the communication system. In this instance, the pulse shaping is distributed equally among the sender filter and receiver filter, and that the amplitude responses of the filters are accordingly point-wise square roots of the filters of the communication system.

4.2.2.1 Sinc Filter

A sinc filter is as well known as a Boxcar filter because its equivalent in frequency-domain is in the form of a rectangular shape. Theoretically speaking, sinc filter is the finest pulse-shaping filter, however, it is hard to realize it precisely. Furthermore, the filter is non-causal with the tails decaying proportionately slowly and in the synchronization viewpoint; this is problematic because any phase error may lead to in sharply increasing ISI.

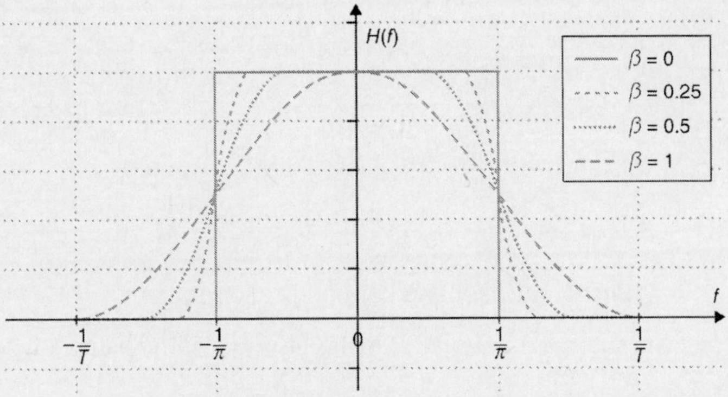

Fig. 4.3: Raised- cosine filter amplitude response for different roll- off factors

4.2.2.2 Raised-cosine Filter

Raised-cosine filter is quite similar to that of a sinc filter, except the fact that for a spectral width that is slightly larger, we can trade off for sidelobes that are smaller. Moreover, a raised-cosine filter is a widely used filter in communication systems as it is practically implementable. Also, the filter has an excess bandwidth that is configurable, and hence, transmission systems can trade for simple filter design or spectral efficiency.

4.2.2.3 Gaussian Filter

The output of this filter yields a pulse that has a shape that resembles a Gaussian function.

4.3 SIGNAL DESIGN FOR ZERO ISI

The main intent of this section is to design an overall pulse shape of $p(t)$ so that we can minimize the ISI, provided the channel impulse response $h(t)$ is given. Based on this objective, the problem statement can be defined as follows.

For the communication system portrayed in the above figure, we need to construct $p(t)$, i.e., the overall shape of the pulse generated by the system such that the receiver can reconstruct exactly the stream of data that was originally applied at the input of the transmitter.

As a result, signalling in a channel that is band-limited will become distortionless, and hence, the pulse-shaping requisite may be viewed as a design related problem of the signal. In the forthcoming section the signal-design method is described, pulses that are overlapping in the transmission system depicted in the figure above will be designed in a way that these pulses never *interfere* with each other at the output of the receiver, at the instant of sampling $t_i = iT_b$. Therefore, the pulses that overlap outside the sampling period mentioned above have no practical importance as long as the data stream that was available originally at the input of the transmitter can be reconstructed. This design process is entrenched in the distortionless transmission criterion devised by Nyquist in the context of the theory of transmission of telegraph signals, which was valid at that time even today.

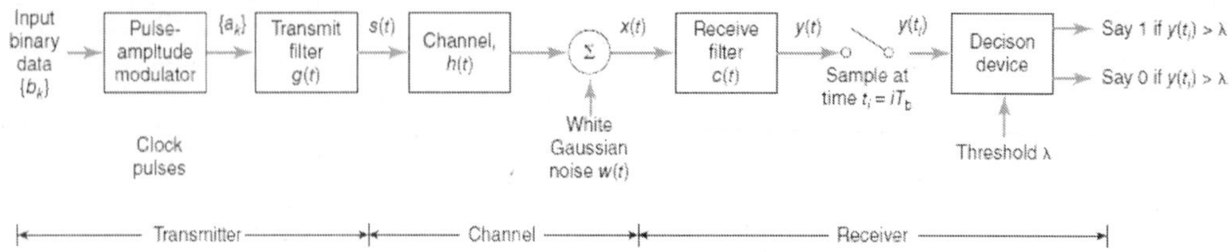

Fig. 4.4: Baseband communication system

The output of the receiver filter $y(t)$ sampled at instant $t_i = iT_b$, with i being integers can be written as

$$y(t_i) = \sum_{k=-\infty}^{\infty} a_k p[(i-k)T_b]$$
$$= a_i + \sum_{\substack{k=-\infty \\ k\neq i}}^{\infty} a_k p[(i-k)T_b] \tag{4.1}$$

In Eq. (4.1), a_i denotes the contribution due to the ith bit that was transmitted and the term inside the summation represents the residual effect imposed on the ith bit by the rest of the transmitted bits while it is decoded. Furthermore, the pulses that occur prior to and after the sampling time t_i cause this residual effect which is termed as ISI. Under the assumption of zero ISI and channel noise, it can be noticed that in Eq. (4.1) term inside the summation will become zero. As a result, Eq. (4.1) reduces to

$$y(t_i) = a_i \tag{4.2}$$

From Eq. (4.2), we may observe that under ideal conditions (zero ISI and channel noise), we can correctly decode ith bit that was transmitted by the transmitter.

Furthermore, from Eq. (4.2), it can be seen that the contribution of the second term in the summation, i.e., $a_k p(iT_b - kT_b)$, the weighted pulse, should be zero for all values of k excluding $k = 1$, if at all if we want to achieve ISI free transmissions over the channel that is band-limited. Alternatively, we may state that the shape of $p(t)$ that denotes the complete pulse should be designed along the lines that it satisfies the following requirement

$$p(iT_b - kT_b) = \begin{cases} 1 & \text{for} \quad i = k \\ 0 & \text{for} \quad i \neq k \end{cases} \tag{4.3}$$

Here, according to the normalization procedure, we may set $p(0)$ equal to unity, i.e., $p(0) = 1$. Now, $p(t)$, the pulse that meets out to the condition defined in Eq. (4.3) that is two-part is referred to as a *Nyquist pulse*, and the condition by its very nature is called as *Nyquist's criterion for distortionless transmission*. On the other hand, there is no distinctive Nyquist pulse; but there exist several shapes of pulse that still fulfil the Nyquist condition of Eq. (4.3). In the subsequent section, we explain two types of Nyquist pulses, with each of them having its own features.

4.4 IDEAL NYQUIST PULSE FOR DISTORTIONLESS BASEBAND DATA TRANSMISSION

In the context of the design viewpoint, it is better to transform (4.3) to the frequency domain. Now, if we consider samples sequence $\{p(nT_b), n = 0, \pm 1, \pm 2, \ldots,\}$ and based on the fact that the sampling process in the time domain yields frequency domain periodicity, we may arrive at the following

$$P_\delta(f) = R_b \sum_{n=-\infty}^{\infty} P(f - nR_b) \tag{4.4}$$

Whereby definition, $R_b = 1/T_b$ gives the *bit rate* with unit bits per second and $P_\delta(f)$ in Eq. (4.4) represents the Fourier transform pertaining to infinite periodic series of delta functions having period T_b with each of the areas being weighted in accordance with the corresponding values of sample of $p(t)$. As a result, $P_\delta(f)$ can be written as

$$P_\delta(f) = \int_{-\infty}^{\infty} \sum_{m=-\infty}^{\infty} \left[p(mT_b)\delta(t - mT_b) \right] \exp(-j2\pi ft)dt \tag{4.5}$$

Let us denote $m = i - k$ as the integer with $i = k$ corresponding to $m = 0$ and, as well $i \neq k$, relating to $m \neq 0$. Consequently, applying the constraints stated in Eq. (4.3) on $p(t)$ sample values that reside in the integral of Eq. (4.5), we may now express (4.5) as follows

$$\begin{aligned} P_\delta(f) &= p(0) \int_{-\infty}^{\infty} \delta(t) \exp(-j2\pi ft)dt \\ &= p(0) \end{aligned} \tag{4.6}$$

Here, the sifting property relating to the delta function is exploited. Furthermore, assuming that we have used normalization, we may write $p(0) = 1$. Therefore, from Eq. (4.4) and (4.6), it follows that zero ISI with respect to frequency-domain condition is satisfied subject to

$$\sum_{n=-\infty}^{\infty} P(f - nR_b) = T_b \tag{4.7}$$

whereby definition, $T_b = 1/R_b$. Hence, in the frequency domain, the *Nyquist criterion* that will result in *transmissions* that is distortionless can be stated as follows. The function $P(f)$ annihilates ISI for the samples considered at instances T_b on the condition that it satisfies Eq. (4.7). Note, furthermore that, the function $P(f)$ represents the overall system according to the following expression.

$$P(f) = G(f)H(f)C(f)$$

To be specific, $P(f)$ in the above expression encompasses $H(f)$, $G(f)$, and $C(f)$ that correspond to the channel, transmit and receive filters, respectively.

4.4.1 Ideal Nyquist Pulse

One of the easiest ways to satisfy Eq. (4.7) is to describe the function $P(f)$ as a *rectangular function*, as given below

$$P(f) = \begin{cases} \frac{1}{2W}, & -W < f < W \\ 0, & |f| > W \end{cases}$$
$$= \frac{1}{2W}\text{rect}\left(\frac{f}{2W}\right) \tag{4.8}$$

In Eq. (4.8), rect(f) denotes the rectangular function corresponding to unit amplitude along with unit support that is centered at $f = 0$. Besides, W that denotes the baseband overall bandwidth of the system can be expressed as follows

$$W = \frac{R_b}{2} = \frac{1}{2T_b} \tag{4.9}$$

From Eq. (4.8), it is obvious that none of the frequencies having an absolute value that exceeds half of the bit rate are required. As a result, and on the basis of the Fourier transform, we have a signal waveform that results in zero ISI that can very well be described by the *sinc function* as given below.

$$p(t) = \frac{\sin(2\pi Wt)}{2\pi Wt}$$
$$= \text{sinc}(2Wt) \tag{4.10}$$

The distinctive bit rate value $R_b = 2W$ is known as the *Nyquist rate* and that W is by and of itself referred to as the *Nyquist bandwidth*. Consequently, $P(t)$ defined in Eq. (4.10) for distortionless transmission is commonly known as the *ideal Nyquist pulse*. Note that, ideal means only half of the bit rate is bandwidth requirement.

Fig. 4.5 depicts the plots of the functions $P(f)$ and $P(t)$. In Fig. 4.5(a), normalized version $P(f)$ is represented for the positive frequencies as well as for negative frequencies. In Fig. 4.5(b), the signalling instants along with the respective centered sampling intervals are illustrated. The $P(t)$ may be viewed as the impulse response corresponding to an ideal low-pass filter having bandwidth W and the passband magnitude equivalent to $1/2W$. It can be observed from the plot that the peak value of $P(t)$ lies at the origin and then passes all the way through zero, specifically at bit duration T_b that is defined by its integer multiples. Therefore, it is evident that if we sample $y(t)$ that corresponds to the waveform of the received signal at time instants $t = 0, \pm T_b, \pm 2T_b, \ldots$, subsequently, the pulses that are described by $a_i p(t - iT_b)$, with $i = 0, \pm 1, \pm 2, \ldots$, and having amplitude a_i are unlikely to interfere among each other. The abovementioned condition is shown in Fig. 4.6 for the binary data 1011010.

The use of an ideal Nyquist pulse yields bandwidth efficiency and solves the zero ISI problem with minimum bandwidth requirement. However, we may end up with two practical difficulties in the context of signal design thereby making it an unattractive solution:

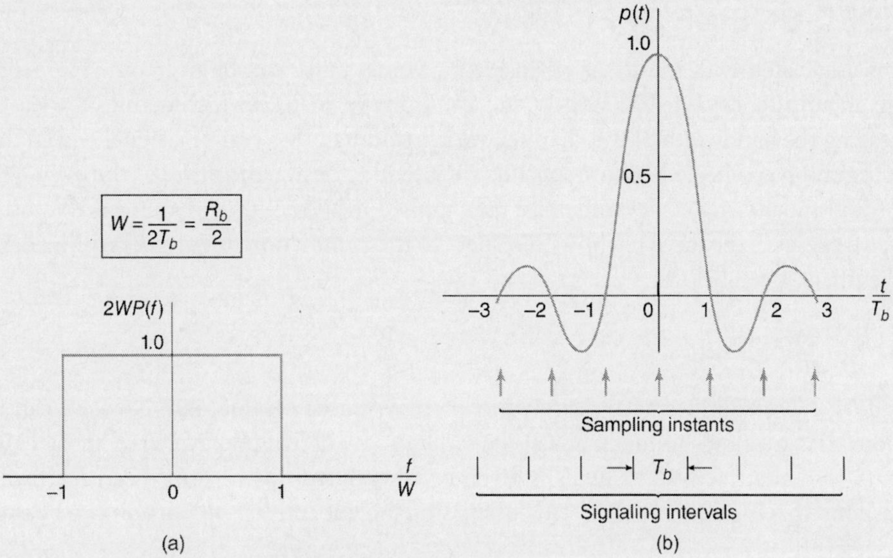

Fig. 4.5: (a) Magnitude response (ideal) of frequency function and (b) shape of an ideal basic pulse

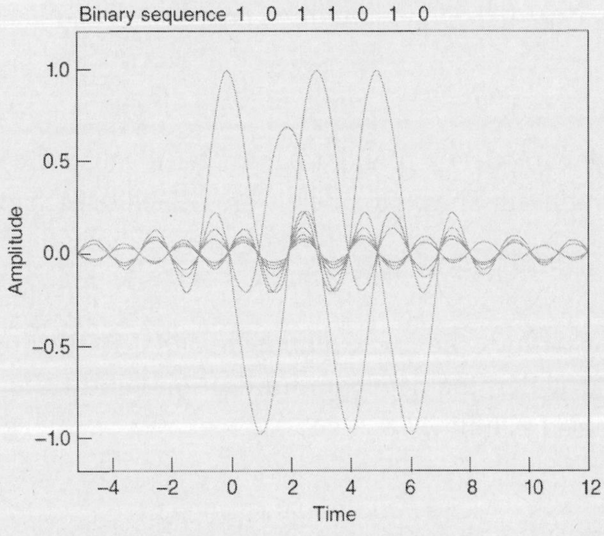

Fig. 4.6: A sequence of sinc pulses that corresponds to the binary data 1011010

(i) Firstly, it needs the $P(f)$ magnitude characteristic to be flat in the range $-W$ to $+W$, as well as zero elsewhere. Nevertheless, this requirement is practically unattainable as there will be rapid transitions at the edges of the band $+W$.

(ii) Secondly, for large $|t|$, $p(t)$ declines at the rate of $1/|t|$. Hence, the decay rate turns out to be slow. Furthermore, this slow rate decay is due to the discontinuity exhibited by the function $P(f)$ at $+W$, and hence, at the receiver, there will almost be no margin of error in the sampling instants.

4.5 RAISED-COSINE SPECTRUM

The problems associated with the usage of an ideal Nyquist pulse can be overcome by increasing the bandwidth to an adjustable value between W and $2W$ from $W = R_b/2$, which defines the least value. In fact, by increasing the bandwidth of the channel, we are trading off to realize a signal design that will be more robust to timing errors. To be more specific, we intend to design the $P(f)$, i.e., the overall frequency response to fulfill more strict constraints than that applied in the context of an ideal Nyquist pulse. In other words, we preserve the three terms with respect to the summation of Eq. (4.7) while restricting the band to $[-W, W]$, as given below

$$P(f)+P(f-2W)+P(f+2W)=\frac{1}{2W},\ -W\leq f\leq W \tag{4.11}$$

In the above expression, we have assumed $R_b = 1/2W$ based on Eq. (4.9). Now, we can formulate many functions that are band-limited while at the same time satisfying Eq. (4.11). A special kind of $P(f)$ that represents several attractive attributes is given by a so-called *raised-cosine* (*RC*) *spectrum* and the frequency response having a sinusoidal form comprising the flat and *roll-off* portion can be shown to be as follows.

$$P(f)=\begin{cases}\dfrac{1}{2W}, & 0\leq|f|<f_1\\ \dfrac{1}{4W}\left\{1+\cos\left[\dfrac{\pi}{2W\alpha}\left(|f|-f_1\right)\right]\right\}, & f_1\leq|f|<2W-f_1\\ 0, & |f|\geq 2W-f_1\end{cases} \tag{4.12}$$

In Eq. (4.12), a new frequency f_1 and a parameter α called the *roll-off* factor that is dimensionless are introduced. These parameters are related to each other by

$$\alpha=1-\frac{f_1}{W} \tag{4.13}$$

The *roll-off* factor defines the *excess bandwidth* that is required compared to the ideal result, W. In particular, the new communication bandwidth that is required for transmission is expressed as follows.

$$\begin{aligned}B_T&=2W-f_1\\&=W(1+\alpha)\end{aligned} \tag{4.14}$$

The normalized frequency response $P(f)$ that is obtained by multiplying $P(f)$ by $2W$, is shown in Fig. 4.7(a) for different values of α, i.e., $\alpha = 0, 0.5$, and 1. From the plot, we can see that $P(f)$ steadily rolls off for $\alpha = 0.5$ or 1 when compared with the $\alpha = 0$, which denotes the ideal Nyquist pulse. Consequently, such a type of filter is easily realizable in practice. Furthermore, as the shape of the roll-off resembles a cosine function, it is termed an *'RC spectrum.'* Also, it is worth mentioning that $P(f)$ demonstrates odd symmetry in connection with W, the Nyquist bandwidth. As a result, $P(f)$ has every possibility to comply with the frequency-domain constraints defined by Eq. (4.7). Now, by taking the inverse Fourier transform of $P(f)$, we may obtain the time domain response $p(t)$, which is expressed as

$$p(t) = \text{sinc}(2Wt)\frac{\cos(2\pi\alpha Wt)}{1-16\alpha^2W^2t^2} \tag{4.15}$$

Fig.4.7(b) demonstrates the response in the time domain for various values of α, i.e., α = 0, 05, and 1.

Note that p(t) is represented as the product of two components namely sinc($2Wt$), which is the first component that characterizes the ideal Nyquist pulse, in addition to a second component that decreases at the rate of $1/|t|^2$ specifically for large $|t|$. Also, the first component will ensure that $p(t)$ crosses the zero at the desirable sampling time defined as $t = iT_b$, where i being a positive and negative integer, while the second component will reduce the pulse tails significantly lower than that attained if an ideal Nyquist pulse is used. Therefore, if such a pulse is used to transmit the binary data, timing errors will have little impact on the transmission. Actually, for α = 1, the roll-off is gradual resulting in oscillatory tails amplitudes of $p(t)$ to be the smallest. As a result, if α increases from 0 to 1, the quantum of ISI due to timing error decreases.

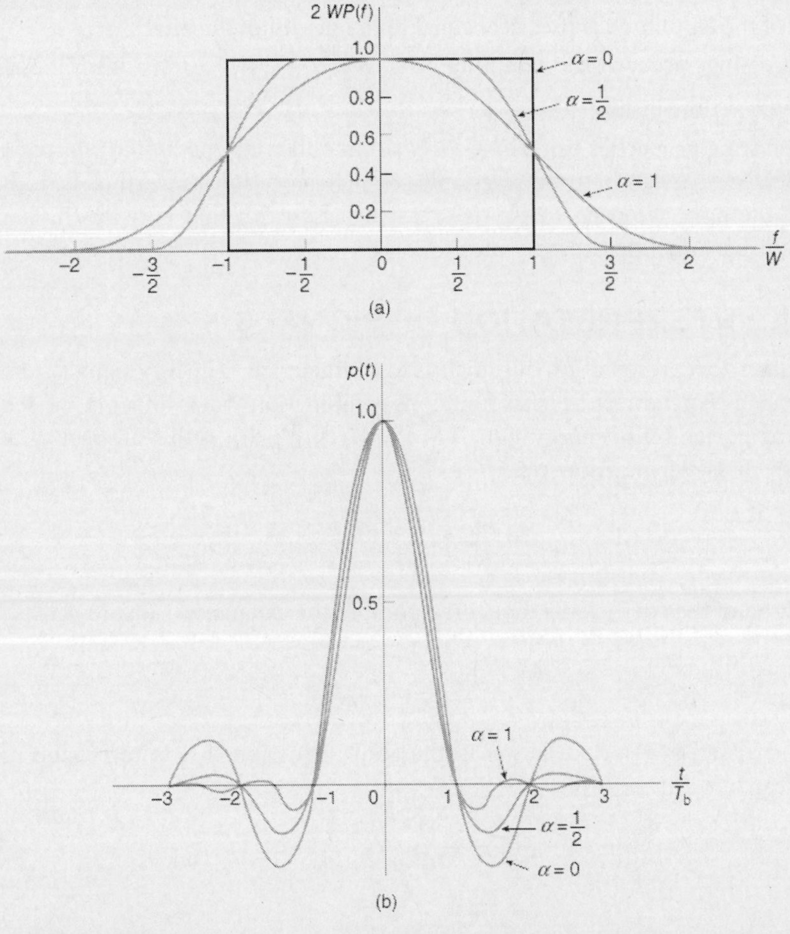

Fig. 4.7: Responses related to various roll- offs: (a) frequency domain response and (b) time domain response.

For a special case with the roll-off α = 1, which designates $f_1 = 0$, the characteristic is termed as *roll-off* of the *full-cosine*. In this context, the frequency domain response described in Eq. (4.12) can be simplified as

$$P(f)=\begin{cases}\frac{1}{4W}\left[1+\cos\left(\frac{\pi f}{2W}\right)\right], & 0<|f|<2W \\ 0, & |f|\geq 2W\end{cases} \tag{4.16}$$

and the corresponding time domain response $p(t)$ can be simplified as

$$p(t)=\frac{\text{sinc}(4Wt)}{1-16W^2t^2} \tag{4.17}$$

The time domain response of Eq. (4.17) presents two attractive properties:

1. The value of $P(t) = 0.5$ at the instants $t=+T_b/2=\pm 1/4W$. This implies the pulse width that is quantified at half of the amplitude is precisely equal to the period of the bit T_b.
2. The typical zero crossings occur at the sampling instants $t=\pm T_b,\ \pm 2T_b,\ldots$. Additionally, zero crossings occur even at time instants

The above two interesting properties can particularly aid in obtaining the *timing information* from the received signal which in turn can be used to establish synchronization. Nevertheless, to have this advantageous property, the price we need to pay is to use a bandwidth that is twice the bandwidth needed for an ideal Nyquist channel that corresponds to α = 0.

4.6 GEOMETRIC INTERPRETATION OF SIGNALS

The quintessence of geometric representation of signals is to characterize *M* energy signals from any set say for example $\{S_i(t)\}$ using *N* orthonormal basis functions combination that is linear in the sense, with $N \leq M$. To be specific, for a given set of energy signals $S_1(t), S_2(t), \ldots, S_M(t)$ that are real-valued, each having a duration of *T* seconds, we may now write

$$s_i(t)=\sum_{j=1}^{N}s_{ij}\phi_j(t),\ \begin{cases}0\leq t\leq T \\ i=1,2,\ldots,M\end{cases} \tag{4.18}$$

whereby the definition of the coefficients that correspond to the expansion is expressed as

$$s_{ij}=\int_0^T s_i(t)\phi_j(t)dt,\ \begin{cases}i=1,2,\ldots,M \\ j=1,2,\ldots,N\end{cases} \tag{4.19}$$

The basis functions $\phi_1(t), \phi_2(t), \ldots, \phi_N(t)$ in the above equation that is real-valued form an *orthonormal set*, which can be written as follows:

$$\int_0^T \phi_i(t)\phi_j(t)dt=\delta_{ij}=\begin{cases}1 & if\ i=j \\ 0 & if\ i\neq j\end{cases} \tag{4.20}$$

Here, δ_{ij} denotes the *Kronecker delta*. In Eq. (4.20), the first constraint says that each of the basis functions is *normalized* so that each of them contains unit energy, while the second constraint says that the basis functions are *orthogonal* relating to each other in the time interval $0 \leq t \leq T$.

For a given *i*, $\{s_{ij}\}_{j=1}^{N}$ that denotes the coefficients set can be considered as an *N-dimensional signal vector*, represented by a vector s_i. Hence, it is worth noting that s_i has a *one-to-one* relation with $S_i(t)$.

If *N* elements corresponding to s_i are given as an input, then directly based on Eq. (4.18), we can make use of the approach illustrated in Fig. 4.8(a), to produce the signal $S_i(t)$, The figure comprises a collection of *N* multipliers with each of the multipliers consisting of its own basis function and a summer. This method shown in Figure 4.8(a) can be considered a *synthesizer*.

On the contrary, if the input signals $S_i(t)$, $i = 1, 2, ..., M$, is given, then directly based on Eq. (4.19), we can make use of the approach illustrated in Fig. 4.8(b) to determine the coefficients $S_{i1}, S_{i2}, \ldots, S_{iN}$. The second method comprises a collection of *N* correlators, also called *product integrators* with each of the correlators having its own basis function and a common input. The method shown in Fig. 4.8(b) can be considered an *analyser*.

Consequently, it can be stated that each of the signals in the signal set $\{S_i(t)\}$ can be entirely calculated with the aid of a signal vector given by

$$\boldsymbol{s}_i = \begin{bmatrix} s_{i1} \\ s_{i2} \\ \vdots \\ s_{iN} \end{bmatrix}, \; i = 1,2,\ldots,M \tag{4.21}$$

Moreover, if the common idea of two and three-dimensional Euclidean spaces is extended to an *N*-dimensional, then the signal vectors set defined as $\{S_i \mid i = 1, 2, ..., M\}$ can be visualized as described in the *N*-dimensional Euclidean space, an equivalent set of *M* points having *N* axes that are mutually perpendicular. These axes are labeled as $\phi_1, \phi_2 \ldots \phi_N$ and this *N*-dimensional space in the Euclidean sense is referred to as the *signal space*. The basic idea behind envisioning geometrically the energy signals, as the one described here, has an insightful importance both in the theoretical and practical sense. Specifically, the geometric representation of the signals gives the mathematical convenience that is satisfying in the conceptual sense. The geometric representation is evinced in Fig. 4.9 for the case of three signals in the signal space that is two-dimensional, i.e., $N = 2$ and $N = 3$.

We may now define the lengths and angles of the vectors in an *N*-dimensional Euclidean space. Usually, the length also known as the *absolute value* or *norm* corresponding to a signal vector S_i is denoted by $\|S_i\|$. Furthermore, for any S_i, the squared length is described by the *dot product or inner product* of S_i among itself, which can be expressed as

$$\begin{aligned} \left\|\boldsymbol{s}_i^2\right\| &= \boldsymbol{s}_i^T \boldsymbol{s}_i \\ &= \sum_{j=1}^{N} s_{ij}^2, \; i = 1,2,\ldots,M \end{aligned} \tag{4.22}$$

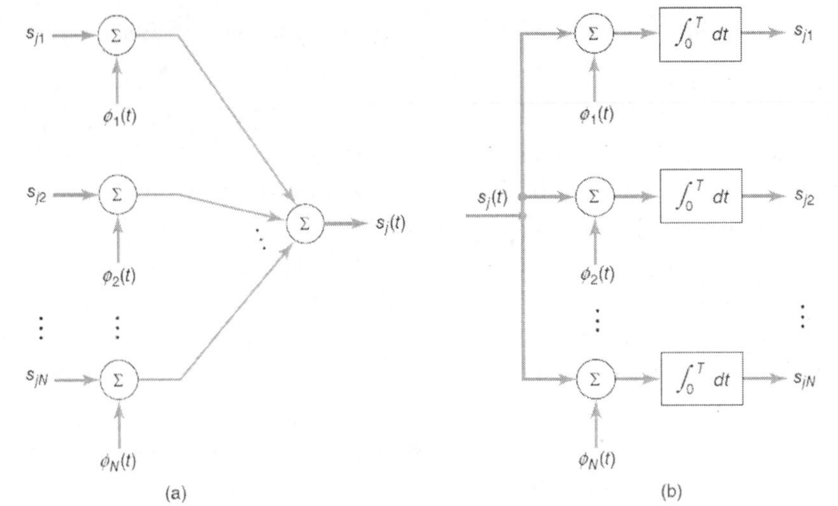

Fig. 4.8: (a) Synthesizer that generates $S_i\ (t)$, (b) Analyser that reconstructs $\{S_i\}$

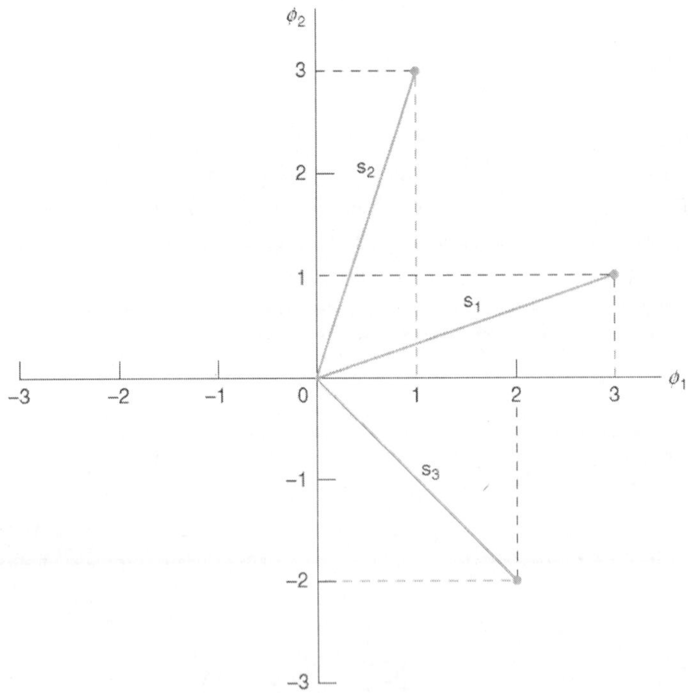

Fig 4.9: Geometric description of signals relating to N = 2 and N = 3

where S_{ij} represents the jth component of S_i and the superscript 'T' represents the transpose of a matrix. By the way, we have an attractive relationship that connects the energy content relating to a signal and its vector description. The energy corresponding to signal $S_i(t)$ with duration T seconds can be defined as follows.

$$E_i = \int_0^T s_i^2(t)dt,\ i = 1,2,\ldots,M \tag{4.23}$$

Hence, using Eq. (4.18) in Eq. (4.23), we have

$$E_i = \int_0^T \left[\sum_{j=1}^{N} s_{ij}\phi_j(t)\right]\left[\sum_{k=1}^{N} s_{ik}\phi_k(t)\right]dt \tag{4.24}$$

As both summation and integration are linear operations, we may now interchange their arrangement, and after rearranging the terms, we arrive at the following.

$$E_i = \sum_{j=1}^{N}\sum_{k=1}^{N} s_{ij}s_{ik}\int_0^T \phi_j(t)\phi_k(t)dt \tag{4.25}$$

Based on the two conditions of Eq. (4.20), and by definition, $\phi_j\ (t)$ forms an orthonormal set. Therefore, Eq. (4.25) simply reduces to

$$\begin{aligned} E_i &= \sum\nolimits_{j=1}^{N} s_{ij}^2 \\ &= \left\|\boldsymbol{s}_i\right\|^2 \end{aligned} \tag{4.26}$$

Accordingly, Eq. (4.22) and (4.26) demonstrate that the energy corresponding to the signal $S_i\ (t)$ is equivalent to the squared length relating to the signal vector $S_i\ (t)$. Suppose, a pair of signals $S_i\ (t)$ and $S_k\ (t)$ are described by their respective signal vectors S_i and S_k, it can be shown that

$$\int_0^T s_i(t)s_k(t)dt = \boldsymbol{s}_i^T\boldsymbol{s}_k \tag{4.27}$$

Eq. (4.27) defines the following:

If $S_i\ (t)$ and $S_k\ (t)$ are energy signals then their inner-product in the time duration $[0, T]$ is same as the inner-product that corresponds to their vector description S_i and S_k. Note, furthermore that the inner-product

$$\boldsymbol{s}_i^T\boldsymbol{s}_k$$

is *invariant* with respect to the basis functions $\left\{\phi_j(t)\right\}_{j=1}^{N}$ choice, wherein, it relies only on the $S_i\ (t)$ and $S_k\ (t)$ signal components that are projected onto each basis function. Besides, one more useful relationship that involves the vector description of $S_i\ (t)$ and $S_k\ (t)$ can be expressed by

$$\begin{aligned} \left\|\boldsymbol{s}_i - \boldsymbol{s}_k\right\|^2 &= \sum_{j=1}^{N}\left(s_{ij} - s_{kj}\right)^2 \\ &= \int_0^T \left(s_i(t) - s_k(t)\right)^2 dt \end{aligned} \tag{4.28}$$

where $||S_i - S_k||$ denotes the *Euclidean distance* d_{ik} connecting the points that are described by the S_i and S_k. In order to completely describe the energy signals in terms of their geometric representation, we need to consider the angle that θ_{ik} subtends between S_i and S_k. Also, the inner-product relating to the vectors S_i and S_k divided by the product of their respective norms gives the cosine of the angle θ_{ik} as shown below

$$\cos(\theta_{ik}) = \frac{\boldsymbol{s}_i^T \boldsymbol{s}_k}{\|\boldsymbol{s}_i\|\|\boldsymbol{s}_k\|} \tag{4.29}$$

If the inner products $\boldsymbol{s}_i^T \boldsymbol{s}_k$ of the vectors S_i and S_k is zero, then these vectors are said to be perpendicular or orthogonal to each other. In this case, $\theta_{ik} = 90°$; intuitively, we may say that this is a satisfying condition.

4.6.1 Gram–Schmidt Orthogonalization Procedure

In the previous section, we have demonstrated the convenience with which we can represent the energy signals geometrically with a suitable example. However, a natural question arises in the sense of how we rationalize it in terms of its mathematical representation. This question is well answered by the *Gram–Schmidt orthogonalization method*, which requires a *complete orthonormal basis functions set*. To continue with the conceptualization of this method, assume a set that has M energy signals designated by $s_1(t), s_2(t), \ldots, s_M(t)$. Suppose, if we start by selecting $s_1(t)$ arbitrarily from the above-mentioned set, then the first basis function can be described as follows

$$\phi_1(t) = \frac{s_1(t)}{\sqrt{E_1}} \tag{4.30}$$

where E_1 in Eq. (4.30) represents the energy associated with the signal $S_i(t)$.

Following this, we may now show

$$\begin{aligned} s_1(t) &= \sqrt{E_1}\phi_1(t) \\ &= s_{11}(t)\phi_1(t) \end{aligned}$$

In the above expression, the coefficient $s_{11} = \sqrt{E_1}$ and $\theta_1(t)$ have unit energy as preferred. Subsequently, if we use signal $S_2(t)$, then the coefficient s_{21} can be expressed as

$$s_{21} = \int_0^T s_2(t)\phi_1(t)dt$$

Now, a new intermediary function is introduced.

$$g_2(t) = s_2(t) - s_{21}\phi_1(t) \tag{4.31}$$

that is orthogonal to the basis function $\phi_1(t)$ in the $0 \le t \le T$ time interval based on the description of S_{21} and by the virtue that $\phi_1(t)$ has unit energy. Following this, the second basis function can be defined as follows.

$$\phi_2(t) = \frac{g_2(t)}{\sqrt{\int_0^T g_2^2(t)dt}} \tag{4.32}$$

Using Eq. (4.31) in Eq. (4.32) and simplifying, we arrive at the following result

$$\phi_2(t) = \frac{s_2(t) - s_{21}\phi_1(t)}{\sqrt{E_2 - s_{21}^2}} \quad (4.33)$$

whereby definition, E_2 denotes the energy corresponding to signal $S_2(t)$. From Eq. (4.32), we can write

$$\int_0^T \phi_2^2(t)dt = 1$$

As a result, Eq. (4.33) results in

$$\int_0^T \phi_1(t)\phi_2(t)dt = 0$$

Hence, we can say that $\phi_1(t)$ and $\phi_2(t)$ establishes themselves as orthonormal pair. Adopting the same procedure, we can generalize as follows

$$g_i(t) = s_i(t) - \sum_{j=1}^{i-1} s_{ij}\phi_j(t) \quad (4.34)$$

The coefficients S_{ij} in Eq. (4.34) can be expressed as

$$s_{ij} = \int_0^T s_i(t)\phi_j(t)dt, \ j = 1,2,\ldots,i-1$$

If we assume $i = 1$, then $g_i(t)$ gets reduced to $S_i(t)$. Furthermore, if $g_i(t)$ is given, then we can describe the basis functions set as

$$\phi_i(t) = \frac{g_i(t)}{\sqrt{\int_0^T g_i^2(t)dt}}, \text{ where } j = 1,2,\ldots,N \quad (4.35)$$

which indeed results in an orthonormal set. For any of the two possibilities stated below, the dimension of N can be less than or equal to M, whereby the definition, M, denotes the number of signals taken into consideration.

If a set of signals $s_1(t), s_2(t), \ldots, s_M(t)$ exhibit linearly independency, then $N = M$.

If a set of signal $s_1(t), s_2(t), \ldots, s_M(t)$ does not exhibit linearly independency, then we have $N < M$.

Furthermore, for $i > N$, the intermediary function $g_i(t)$ takes the value zero.

EXERCISES

Two mark questions

1. What is pulse shaping?
2. Name three common pulse-shaping filters.
3. Define bit rate and give its expression.

4. What is the Nyquist criterion condition to achieve zero ISI?
5. Give the expression for an ideal Nyquist pulse.
6. Define the Nyquist bandwidth and Nyquist rate.
7. Give the expression for excess bandwidth for a raised cosine filter.
8. Define the orthonormal basis function.
9. Define Kronecker delta function.

Five mark questions

1. Explain different types of line coding schemes with an example.
2. Explain three most commonly employed pulse-shaping filters.
3. Explain the Gramm–Schmidt Orthogonalization procedure.

Ten mark questions

1. Neatly sketch and explain the baseband communication system, which is used to achieve zero ISI.
2. Write detailed notes with diagram on line-coding techniques.

Numerical Problems

1. A computer puts out binary data at the rate of 56 kbps. The computer output is transmitted using a baseband binary PAM system that is designed to have a raised-cosine pulse spectrum. Determine the transmission bandwidth required for each of the following roll-off factors:

   (a) 0.25 (b) 0.5

   (c) 0.75 (d) 1

Solution:

If data rate is 56 kbps, then $B_0 = 28$ kHz.

$$B_T = 28\text{k} \times (1 + 0.25) = 35\,\text{kHz}$$

$$B_T = 28\text{k} \times (1 + 0.5) = 42\,\text{kHz}$$

$$B_T = 28\text{k} \times (1 + 0.75) = 49\,\text{kHz}$$

$$B_T = 28\text{k} \times (1 + 1) = 56\,\text{kHz}$$

2. A binary PAM wave is to be transmitted over a low-pass channel with bandwidth of 75 kHz. The bit duration is 10μs. Find a raised-cosine pulse spectrum that satisfies these requirements.

Solution:

Given that $B_T = 75$ kHz and $B_0 = 50$ kHz. We know that the raised cosine pulse bandwidth is,

$$B_T = 2B_0 - f_1$$

$$f_1 = 2B_0 - B_T$$

$$f_1 = 2 \times 50\text{k} - 75\text{k}$$

$$f_1 = 25\,\text{kHz}$$

The roll-off factor is given as:

$$\alpha = 1 - f_1 / B_T$$

$$\alpha = 1 - 25\,\mathrm{K/50K}$$

$$\alpha = 0.5$$

The design parameters of the required raised-cosine pulse spectrum are f_1 = 25kHz and α = 0.5.

3. Consider a channel with bandwidth 3.0 kHz, which is available for data transmission using binary PAM. Plot the permissible bit (signalling) rate ($1/T_b$) versus the excess bandwidth (f_v) assuming that the roll-off factor (α) varies from zero to unity, and that the criterion for zero ISI is satisfied.

Solution:

The transmission bandwidth and the excess bandwidth are given as $B_T = B_0(1 + \alpha)$ and $f_v = \alpha B_0$, respectively. So, we have

$$B_T = B_0 + f_v$$

where, $B_0 = 1/(2T_b)$. Thus, the signalling rate as a function of excess bandwidth is given as,

$$1/T_b = 2(B_T - f_v)$$

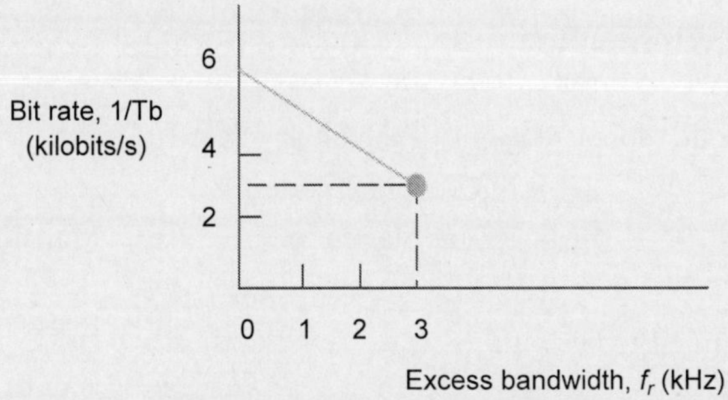

We see that the bit rate decreases linearly with respect to excess bandwidth for a fixed channel bandwidth. In the figure, we see that for a specific bandwidth B_T = 3 kHz, we see that excess bandwidth attains largest value when the value of roll-off factor is unity.

4. You are given a channel of bandwidth 3.0 kHz. The requirement is to transmit data over the channel at the rate of 4.5 kbps using binary PAM.
   (a) What is the maximum roll-off factor in the raised-cosine pulse spectrum that can accommodate this data transmission?
   (b) What is the corresponding excess bandwidth?

Solution:

Given B_T = 3kHz and $1/T_b$ = 4.5 kbps

(a) $B_0 = 1/(2T_b) = 4.5\,K/2 = 2.25$ kHz

  (a) Roll-off factor, $\alpha = (B_T / B_0) - 1$

    1. $\alpha = (3\mathrm{K} / 2.25\mathrm{k}) - 1 = 0.33$

(b) Excess bandwidth, $f_v = \alpha\, B_0$

$$f_v = (1/3) \times 2.25\ \mathrm{k} = 0.75\ \mathrm{kHz}$$

5. In a certain telemetry system, there are eight analog measurements, each of bandwidth 2 kHz. Samples of these signals are time-division multiplexed, quantized, and binary coded. The error in sample amplitudes cannot be greater than 1% of the peak amplitude.
   (a) Determine the number of quantization levels.
   (b) Find the transmission bandwidth Br if Nyquist criterion pulses with roll-off factor $r = 0.2$ are used. The sampling rate must be at least 25% above the Nyquist rate.

Solution:

Quantization error, $\frac{\Delta v}{2} = \frac{m_p}{L} \leq 0.01 m_p \Rightarrow L \geq 100$

Since L should be power of 2, We choose $L = 128 = 2^7$

The Nyquist rate for each signal, $f_{NS} = 2 \times 2000 = 4\,\text{kHz}$

Sampling rate for each signal, $f_S = 1.25 \times 4\text{k} = 5\,\text{kHz}$

Sampling rate for 8 signals, $F_s = 8 \times 5\text{k} = 40\,\text{kHz}$

Bit rate for 7 bits, $R_b = 7 \times 40\text{k} = 280$ kbps

Transmission bandwidth, $B_T = \frac{(1+\alpha)R_b}{2} = \frac{1.2x\,280\,\text{k}}{2} = 168\,\text{kHz}$

6. Find the following line code representations of the binary data 10110001.
   (a) NRZ unipolar code
   (b) NRZ bipolar code
   (c) RZ unipolar code
   (d) RZ bipolar code
   (e) Manchester code

Solution:

If the amplitudes are assumed to be 1 and −1 (i.e. $-A = 1$ and $-A = -1$)

(a) NRZ unipolar code

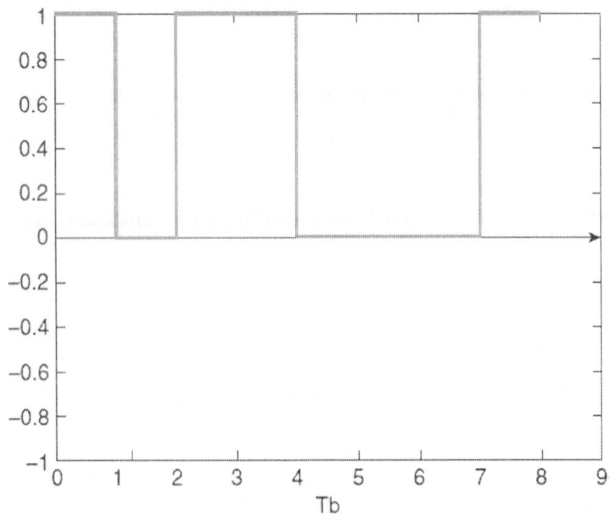

(b) NRZ bipolar code

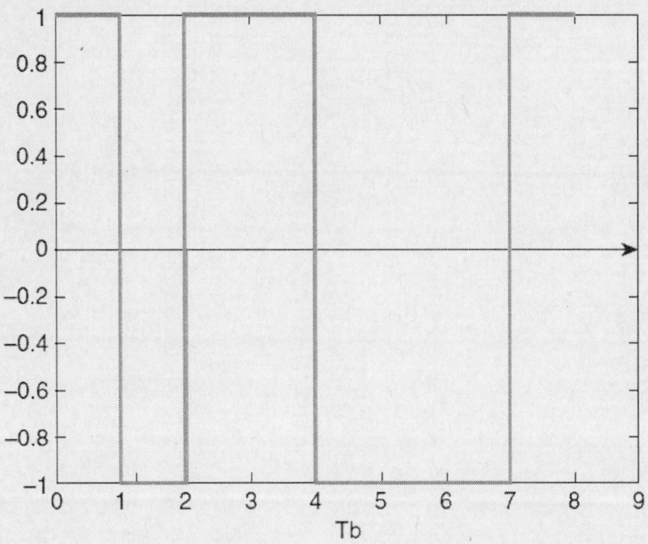

(c) RZ unipolar code

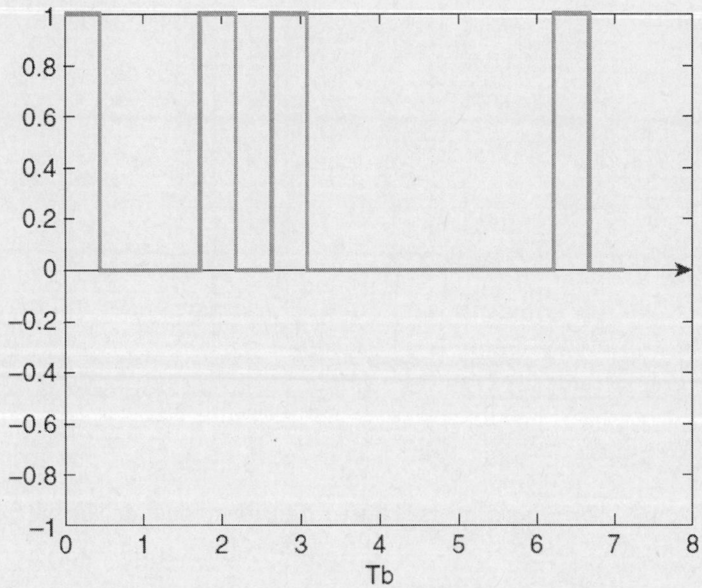

(d) RZ bipolar code

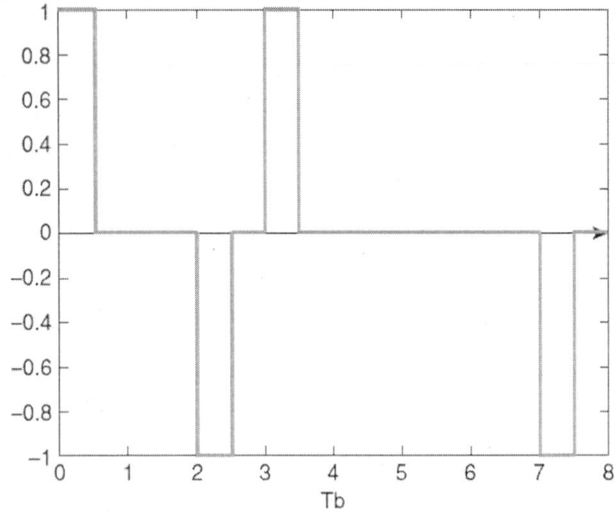

(e) Manchester code

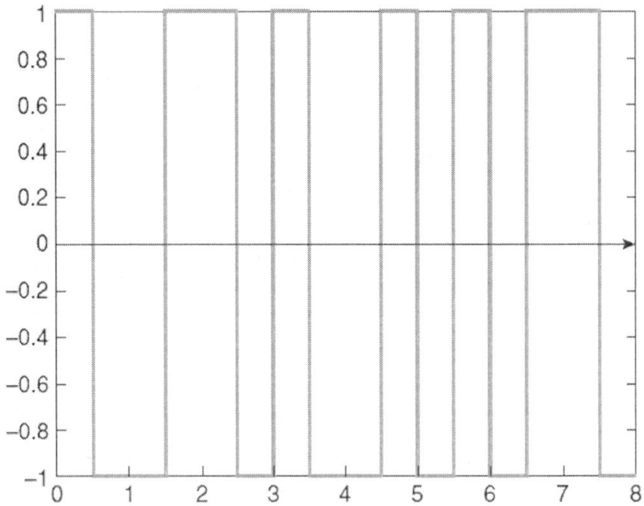

7. Determine the Nyquist pulse whose inverse Fourier transform is defined by the frequency function *P*(*f*) as,

$$P(f)=\begin{cases}\dfrac{1}{2B_0}, & 0<|f|f_1\\[2ex] \dfrac{1}{4B_0}\left\{1+\cos\left[\dfrac{\pi(|f|-f_1)}{2B_0-2f_1}\right]\right\}, & f_1<f<2B_0-f_1\\[2ex] 0, & |f|>2B_0-f_1\end{cases}$$

Solution:

Since *P(f)* is an even real-valued function, its inverse Fourier transform can be simplified as the below equation.

$$p(t) = 2\int_0^{\infty} P(f)\cos(2\pi ft)df$$

Considering *P(f)*,

$$P(f) = \begin{cases} \dfrac{1}{2B_0}, & 0 < |f| f_1 \\ \dfrac{1}{4B_0}\left\{1+\cos\left[\dfrac{\pi(|f|-f_1)}{2B_0-2f_1}\right]\right\}, & f_1 < f < 2B_0 - f_1 \\ 0, & |f| > 2B_0 - f_1 \end{cases}$$

Substituting the above equation in the inverse Fourier transform formula, we have

$$p(t) = \frac{1}{B_0}\int_0^{f_1}\cos(2\pi ft)df + \frac{1}{2B_0}\int_{f_1}^{2B_0-f_1}\left[1+\cos\left(\frac{\pi(f-f_1)}{2B_0\alpha}\right)\right]\cos(2\pi ft)df$$

$$= \left[\frac{\sin(2\pi ft)}{2\pi B_0 t}\right] + \left[\frac{\sin(2\pi ft)}{4\pi B_0 t}\right]_{f_1}^{2B_0-f_1}$$

$$+\frac{1}{4}B_0\left[\frac{\sin\left(2\pi ft + \dfrac{\pi(f-f_1)}{2B_0\alpha}\right)}{2\pi t + \pi/2B_0\alpha}\right]_{f_1}^{2B_0-f_1} + \frac{1}{4B_0}\left[\frac{\sin\left(2\pi ft - \dfrac{\pi(f-f_1)}{2B_0}\right)}{2\pi t + \pi/2B_0\alpha}\right]_{f_1}^{2B_0-f_1}$$

$$= \frac{\sin(2\pi f_1 t)}{4\pi B_0 \mathrm{t}} + \frac{\sin[2\pi t(2B_0 - f_1)]}{4\pi B_0 \mathrm{t}}$$

$$-\frac{1}{4B_0}\frac{\sin(2\pi f_1 t) + \sin[2\pi t(2B_0 - f_1)]}{2\pi t - \pi/2B_0\alpha} + \frac{\sin(2\pi f_1 t) + \sin[2\pi t(2B_0 - f_1 t)]}{2\pi t - \pi/2B_0\alpha}$$

$$= \frac{1}{B_0}\left[\sin(2\pi f_1 t) + \sin[2\pi t(2B_0 - f_1)]\right]\left[\frac{1}{4Wt} - \frac{\pi t}{(2\pi t)^2 - (\pi/2B_0\alpha)^2}\right]$$

$$= \frac{1}{B_0}\left[\sin(2\pi B_0 t)\cos(2\pi\alpha B_0)\right]\left[\frac{-\pi/(2B_0\alpha)^2}{4\pi t\left[(2\pi t)^2 - \pi/(2B_0\alpha)^2\right]}\right]$$

$$= \mathrm{sinc}(2B_0 t)\cos(2\pi\alpha B_0 t)\left[\frac{1}{1-16\alpha^2 B_0^2 t^2}\right]$$

8. This problem follows up on the pulse-shaping criterion for zero ISI. The pulse spectra shown in parts (a) and (b) of the following figure are two examples that can also satisfy the pulse-shaping criterion.
   (a) Derive the condition that the bandwidth in Fig. (a) Must satisfy for the requirement of zero ISI to be satisfied for binary PAM.
   (b) Repeat the problem for the pulse spectrum of Fig. (b).

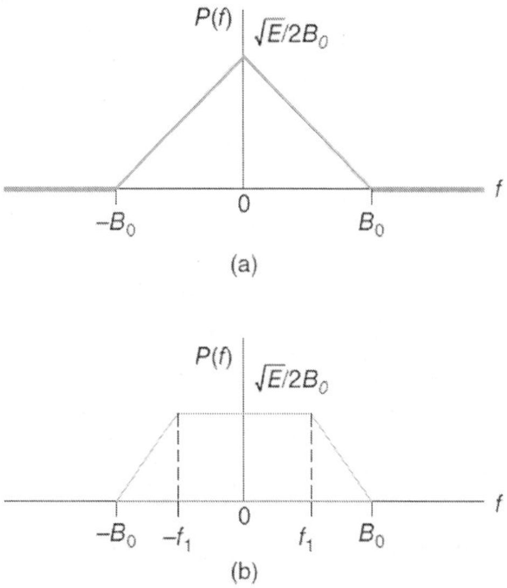

Solution:

The pulse-shaping criterion for zero ISI is given by the following equation where $R_b = 1/T_b$.

$\sum_{n=-\infty}^{\infty} P(f - nR_b) = T_b$

The pulse-shaping spectrum of Fig. (a) is given as,

$$P(f) = \begin{cases} \sqrt{E}/(2B_0) & \text{for } f = 0 \\ \dfrac{\sqrt{E}}{2B_0}\left(1 - \dfrac{f}{B_0}\right) & \text{for } 0 < f < B_0 \\ 0 & \text{for } f = B_0 \end{cases}$$

Substituting the above in the pulse-shaping criterion for zero ISI, we have

$$\frac{1}{T_b} = \frac{B_0}{2}$$

Or equivalently,

$$B_0 = \frac{1}{T_b}$$

The pulse-shaping spectrum of Fig. (b) is given as,

$$P(f) = \begin{cases} \sqrt{E}/(2B_0) & \text{for } 0 \le |f| < f_1 \\ \dfrac{\sqrt{E}}{2B_0}\left(1 - \dfrac{f - f_1}{B_0 - f_1}\right) & \text{for } f_1 < f < B_0 \\ 0 & \text{for } f > B_0 \end{cases}$$

Substituting the above in the pulse-shaping criterion for zero ISI, we have

$$\frac{1}{T} = \frac{1}{2}(f_1 + B_0)$$

Or equivalently,

$$B_0 = \frac{2}{T} - f_1$$

9. An analog signal is sampled, quantized and encoded into a binary PCM wave. The number of representation levels used is 128. A synchronizing pulse is added at the end of each code word. The resulting PCM signal is transmitted over a channel of bandwidth 13 kHz using a quaternary PAM system with a raised-cosine pulse spectrum. The roll-off factor is unity.
   (a) Find the rate (in bits per second) at which information is transmitted through the channel.
   (b) Find the rate at which the analog signal is sampled. What is the maximum possible value for the highest frequency component of the analog signal?

Solution:

The code word consists of $\log_2(128) = 7$ bits. With an additional bit added for synchronization, the overall code word consists of 8 bits. The method of transmission is quaternary (i.e. 4-level PAM) and the roll-off factor is

(a) For binary PAM, the data rate is,

$$R_b = \frac{2B_T}{1 + \alpha} = \frac{2 \times 13}{1 + 1} = 13 \text{ kbps}$$

For quaternary PAM, the data rate is,

$$R = \log_2(L) \times R_b = \log_2(4) \times 13 = 26 \text{ kbps}$$

(b) Each element of the overall code word must fit into the bit duration,

$$T_b = \frac{1}{13 \times 10^3} = 77\ \mu\text{s}$$

With each code word containing 8 bits, the total period is

$$T_s = 8 \times 77 = 616\ \mu\text{s}$$

The sampling rate applied to the analog signal is,

$$R_s = \frac{1}{T_s} = \frac{1}{616} = 162 \text{ kHz}$$

The highest frequency component of the analog signal is,

$$f_{max} = \frac{R_s}{2} = \frac{162 \text{ K}}{2} = 81 \text{ kHz}$$

10. Three signals $S_1(t)$, $S_2(t)$, and $S_3(t)$ are as shown in the figure. Apply the Gram–Schmidt procedure to obtain and orthonormal basis functions for the signals. Express the signals $S_1(t)$, $S_2(t)$, and $S_3(t)$ in terms of orthonormal basis functions. Also give the signal constellation diagram.

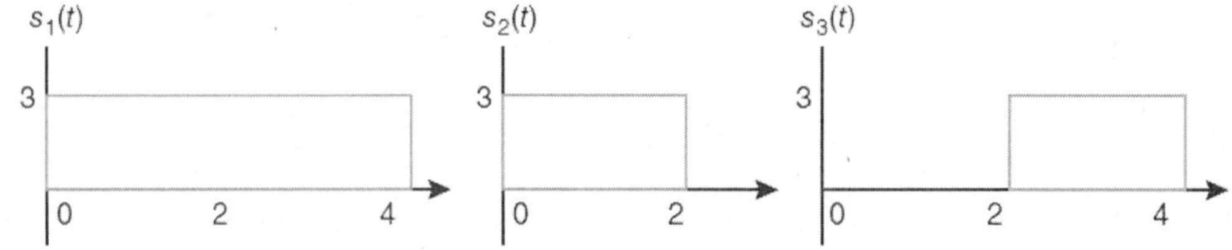

Solution:

We know that,

$$s_1(t) = s_{11}\phi_1(t)$$

$$\int_0^4 s_1^2(t)dt = 9 \times 4 = 36 = s_{11}^2$$

$$\phi_1(t) = \frac{s_1(\text{t})}{s_{11}} = \frac{s_1(t)}{6} = \frac{1}{2}$$

$$s_2(t) = s_{21}\phi_1(t) + s_{22}\phi_2$$

$$\int_0^T s_2 t\phi_1(t)dt = \int_0^2 3 \times \frac{1}{2} dt$$

$$= \frac{3}{2} \times 2 = 3 = s_{21}$$

$$s_{22}\phi_2(t) = s_2(t) - s_{21}\phi_1(t)$$

$$\int_0^T s_{22}^2\phi_2^2(t)dt = s_{22}^2 = \int_0^T (s_2(t) - 3\phi_1(t))^2 dt$$

$$s_{22}^2 = \int_0^2 s_2^2(t)dt + \int_0^4 9\phi_1^2(t)dt - 2\int_0^2 3s_2(t)\phi_1(t)dt = 9$$

$$s_{22} = 3$$

$$\phi_2(\text{t}) = \frac{1}{3}\left[s_2(t) - s_{22}\phi_1(t)\right] = \frac{1}{3}\left[3 - 3\frac{1}{2}\right] = \frac{1}{2}$$

From the figure, it is observed that $S_3(t) = S_1(t) - S_2(t)$. Expressing the signals in terms of basis functions,

$$s_1(t) = \phi_1(t) \times 6$$

$$s_2(t) = \phi_1(t) \times 6 + \phi_2(t) \times 3$$

$$s_3(t) = 3 \times \phi_1(t) - 3 \times \phi_2(t)$$

KNOW MORE

Erhard Schmidt is well known for his work namely the Gram-Schmidt orthogonalization process, which considers the basis of a space and forms an orthogonal one from it. Under Hilbert's supervision, Erhard Schmidt received his doctorate from the University of Göttingen in 1905 for his doctoral thesis. His dissertation focussed mainly on integral equations. The key ideas of his dissertation appeared in Schmidt's 1907 paper. After receiving his doctorate, Erhard Schmidt moved to Bonn. In 1906, at Bonn, Erhard Schmidt was awarded his habitation habilitation (degree). Before Schmidt was awarded with a professorship at the University of Berlin in 1917, he held various positions in Zürich, Erlangen, and Breslau. Schmidt's main area of interest was in integral equations and Hilbert space. Around the year 1905, Erhard Schmidt combined several ideas from Hilbert on integral equations and introduced the Hilbert space concept. He contributed a lot in bringing Berlin to play a leading role in applied mathematics.

REFERENCES AND SUGGESTED READINGS

1. Wayne Tomasi: *Electronic Communication System*, Pearson Education, 5th edition, 2008.
2. Taub and Schilling: *Principles of Communication Systems*, Tata McGraw-Hill, New Delhi, 1995.
3. Simon Haykin: *Communication Systems*, Wiley Eastern, 4th edition, 2001.

Reference for further reading

1. http://digimat.in/nptel/courses/video/108101113/L39.html

2. http://www.digimat.in/nptel/courses/video/108108109/L21.html

5 Spread-Spectrum Modulation

UNIT SPECIFICS

Through this unit, we shall discuss the following aspects:

- Fundamentals of spread-spectrum with its classification and features
- Pseudo-noise (PN) sequences
- Maximum length sequences and selection of a maximum length sequence
- Direct sequence spread spectrum with coherent BPSK: transmitter and receiver modules
- Performance of DSSS systems: the processing gain and probability of error of DS/BPSK systems
- Frequency hop spread-spectrum (FHSS) signals
- Transmitter and receiver of frequency hopping (FH)/M-ary frequency shift keying (MFSK)
- Fast frequency hopping and comparison between slow and fast FH
- Applications of spread-spectrum modulation
- Code division multiple access (CDMA)

RATIONALE

This unit on spread-spectrum (SS) modulation begins with the introduction and classification of SS techniques. The salient characteristics of SS signals are listed in the beginning. The fundamental description about pseudo-noise (PN) sequences are illustrated in detail. Later, the theory behind maximum length sequences and its design are discussed in detail. The transmitter and receiver of DSSS with coherent BPSK have been explained in detail with neat block diagrams. The processing gain and probability of error of DS/BPSK formulae are given. The concepts of frequency hop spread-spectrum (FHSS) with slow frequency hopping (FH) are elaborated. The transmitter and receiver of FH/M-ary frequency shift keying (MFSK) are elucidated.

The chip rate expressions are provided to understand the hopping concept. The details of fast FH with AWGN channel are also presented. The fundamental comparison between slow and fast FH is shown. The applications of the SS are listed completely. Finally, the major application of SS, i.e. CDMA has been explained with necessary information.

PRE-REQUISITES

BPSK and MFSK modulation techniques

UNIT OUTCOMES

List of outcomes of this unit is as follows:

U5-O1: Understand the fundamentals of SS signals with details about classification and characteristics.

U5-O2: Explain concept of PN sequences with illustrations on maximum length sequences and its selection.

U5-O3: Discuss the transmitter and receiver modules of DSSS with coherent BPSK.

U5-O4: Analyse the processing gain and probability of error of a DS/BPSK system.

U5-O5: Study the FHSS techniques, applications of SS and CDMA techniques.

Unit-5 Outcome	EXPECTED MAPPING WITH COURSE OUTCOMES (1 - Weak Correlation; 2 - Medium Correlation; 3 - Strong Correlation)					
	CO-1	CO-2	CO-3	CO-4	CO-5	CO-6
U5-O1	1	-	-	-	3	1
U5-O2	1	-	-	-	3	1
U5-O3	1	2	-	-	3	1
U5-O4	1	-	-	-	3	1
U6-O5	1	2	-	-	3	1

MODULE 5: SPREAD-SPECTRUM MODULATION

5.1 INTRODUCTION

Spread spectrum, a technique widely used during World War II, was developed for secure communication. Its main goal was to send information safely, preventing enemy interference. Enemies could disrupt communication by interfering with the transmitted signal. In the spread spectrum, the signal sent from the transmitter is spread out in the frequency domain using special codes and then narrowed down again at the receiver, making it difficult for adversaries to intercept. The spread signal covers a much broader range of frequencies than the original message signal, which is why it is called 'spread spectrum.'

The spread spectrum has the following applications:

1. Wireless local area networks (LANs)
2. CDMA radios
3. Cordless phones: Many companies introduced a spread-spectrum concept in cordless phones for better security, immunity to noise and long ranger.

5.1.1 Classification of Spread-Spectrum Systems

Spread-spectrum modulation techniques are classified depending upon their operating concept or modulation techniques.

(A) Classification Based on the Operating Concept

The spread-spectrum system falls on two types: averaging and avoidance types.

(i) **Averaging systems:**

It is a technique where the transmitted signal is distributed across a range of frequencies. Receiver collects and averages these signals to enhance communication performance by reducing the impact of noise and interference. This helps to ensure more robust and secure data transmission.

(ii) **Avoidance systems:**

It is a technique where the transmitter and receiver quickly switch frequencies in a coordinated manner to minimize the chances of signal interception and to operate in a crowded or hostile radio frequency environment effectively. This method enhances communication security and reliability.

(B) Classification Based on Modulation Techniques

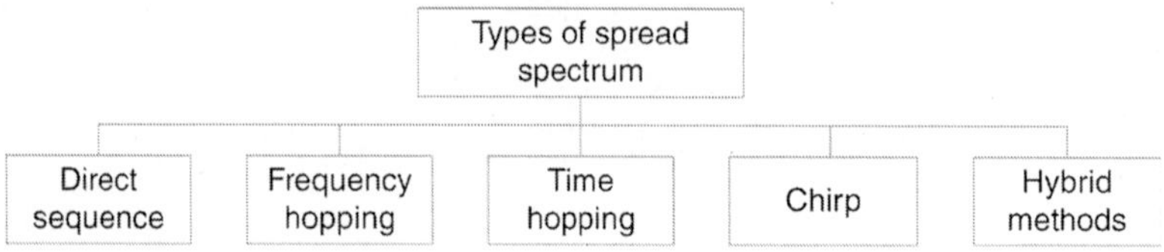

In the abovementioned modulation techniques, direct sequence is an averaging type system and remaining techniques are avoidance type systems. In this unit, we will study the DSSS and FHSS techniques in detail.

(i) **Direct sequence spread spectrum**

It is a communication method that multiplies the input signal by a high-rate pseudo-random bit sequence (chipping code) to spread it out. This spreads the signal over a wide bandwidth, providing resistance to interference and jamming. This spread signal is again modulated using modulation schemes such as binary phase shift keying (BPSK) or quadrature phase shift keying (QPSK), allowing for robust and secure data transmission in noisy environments.

(ii) **Frequency hop spread spectrum**

It is a communication technique where the transmitter and receiver periodically switch among a set of predefined frequencies. This hopping pattern is synchronized, making it resistant to interference and jamming. FHSS often uses modulation schemes such as frequency shift keying (FSK) to encode and transmit data on each frequency hop.

For both direct sequence and frequency hop spread-spectrum modulation techniques, we use pseudo-random or PN sequence. It is a random spreading code. Thus, PN sequence is essential for the operation of the spread-spectrum modulation. In this unit, we will discuss in detail about PN sequences.

5.1.2 Features of Spread Spectrum System

The followings are the salient characteristics of spread-spectrum systems:

Anti-jamming capability: This denotes the spread-spectrum technique's capacity to function effectively in the presence of interference. It is measured as the proportion of average signal power (P) to average interference power (I). This is also referred to as jamming margin.

Ranging functionality: The DSSS method is employed to determine the distance between a spacecraft and an earth station, commonly known as ranging.

Multiple access capability: The SS is used in CDMA systems, wherein numerous end users concurrently communicate over the same channel. Each user is allotted a distinct code. As the codes are unique, different user signals will not undergo interference.

Message security: In SS communication, messages are transmitted over a broad bandwidth employing PN sequences. At the receiver, the original message is extracted using the same PN sequence that was employed at the transmitter. Only authorized recipients possess knowledge of the correct PN sequence, ensuring their ability to retrieve the message accurately.

5.2 PN SEQUENCES

The PN sequence is like a binary code that seems random but actually repeats after a while. It is generated using a special circuit called a feedback shift register with flip-flops (these are like tiny memory cells that store bits 1 or 0). When a clock pulse is given, the bits from one flip-flop moves to the subsequent flip-flop, and this data, along with what's in all the flip-flops, goes into a logic circuit. The output of the logic circuit serves as the first flip-flop's input in the shift register.

The pattern we are interested in comes out of the final flip-flop. Every time the clock ticks, the flip-flop's state moves to the next one, and the logic circuit's output shifts into the first flip-flop. This repeating pattern of the PN sequence shows up again after a certain number of bits, specifically 2^m bits. This 2^m gives the period of the PN sequence.

Now, there's a trick with the logic circuit – it is usually something called a mod-2 adder. If the shift register somehow gets stuck in a state when everything is zeros, the output sequence will be all zeros too, which is not very useful. So, to avoid this, we make sure the shift register never gets into this all-zeros state. For an m-state feedback register, the overall number of states are $(2^m - 1)$. As a result, the output PN sequence also repeats, but it has a period of $(2^m - 1)$ bit instead of 2^m bits.

5.2.1 Maximum Length Sequences

Maximum length sequences are those PN sequences that are formed with a length of $(2^m - 1)$.

5.2.1.1 Properties of Maximum Length Sequences

Balance property: In actual random binary sequence, bits take values as 1's and 0's in equiprobable manner. The total number of 1's is always one over the total number of 0's in each period of a maximum duration sequence.

Run property: A 'run' is a string of similar symbols, either 1's or 0's, which appear repeatedly in each cycle of the sequence. The length of this specific subsequence is equal to the total length of a run. The total number of runs in the pattern is equal to $(2^m - 1)$ when the maximum length sequence is produced with a feedback shift register having length 'm'.

Correlation property: The autocorrelation function has a recurring pattern and has values in binary for maximum length sequences. It quantifies how similar two sequences are when they are shifted by a specific amount of time.

5.2.1.2 Length and Period of a Maximum Length Sequence

The period (in digits) of the maximum length sequence (N) is given as $N = (2^m - 1)$. The frequency at which the bits of a maximum length sequence appear is referred to as the chip rate (R_c), measured in chips per second. Then, the duration of every bit is $T_c = \frac{1}{\text{Chip Rate}} = \frac{1}{R_c}$

Hence, the time period of the sequence is determined as follows $T_b = N\,T_c$. Here, T_b is the time period of the sequence in seconds, whereas N is the length of the sequence in binary digits.

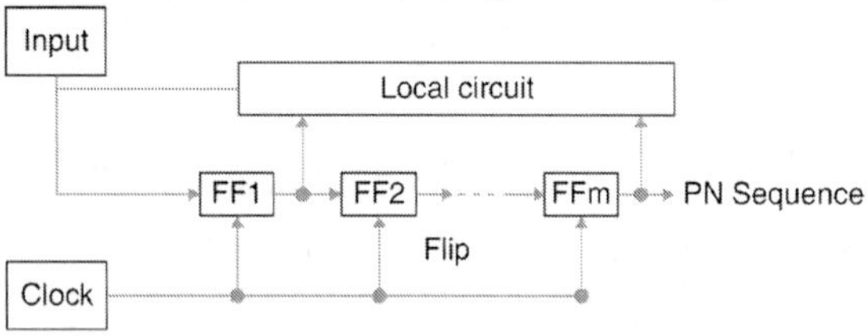

Fig. 5.1: Generation of the PN sequence

5.2.2 Selection of a Maximum Length Sequence

Here, we would see difficulty in generating these sequences. In Fig 5.1, we have shown that a PN sequence can be generated with the help of shift register and feedback logic. Our point of discussion is how to select the feedback logic for a particular length. From the theory of error control coding, some of the feedback logics are developed. These logics are available as standard tables. A table for 8 stage shift registers is given below:

In Table 5.1, we observe that for length $m = 5$, there are three combinations of feedback taps. i.e., (5, 2), (5, 4, 3, 2) and (5, 4, 2, 1).

(i) **Scheme for (5, 2)**

Fig. 5.2 shows the scheme for (5, 2). Here, (5, 2) means, outputs of flip-flops 5 and 2 are added as per mod-2 and given to first flip-flop.

Table 5.1: Feedback taps for shift registers up to 8 stages

Stage of shift register (m)	Feedback taps
2	(2,1)
3	(3,1)
4	(4,1)
5	(5,2);(5,4,3,2);(5,4,2,1)
6	(6,1);(6,5,2,1);(6,5,3,2)
7	(7,1);(7,3);(7,3,2,1);(7,4,3,2);(7,6,4,2);(7,6,3,1);(7,6,5,2);(7,6,5,4,2,1);(7,5,4,3,2,1)
8	(8,4,3,2);(8,6,5,3);(8,6,5,2);(8,5,3,1);(8,6,5,1);(8,7,6,1);(8,7,6,5,2,1);(8,6,4,3,2,21)

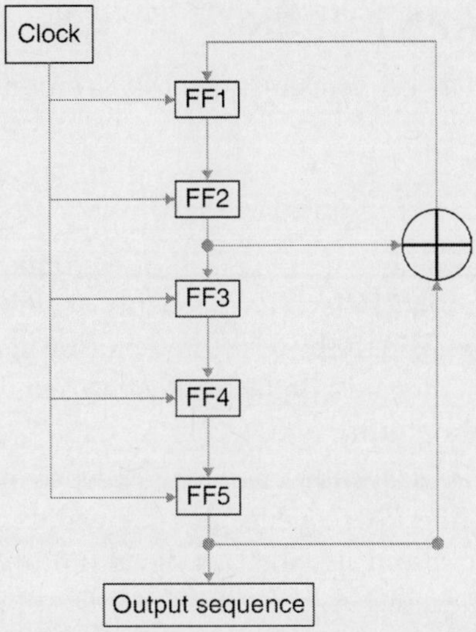

Fig. 5.2: Feedback shift registers with (5, 2) connection

(ii) **Scheme for (5, 4, 3, 2)**

Fig. 5.3 shows the scheme for (5, 4, 3, 2) to get a maximum length sequence. As shown in Fig. 5.3, the outputs of flip-flop 5, 4, 3 and 2 are mod-2 added and given to the first flip-flop.

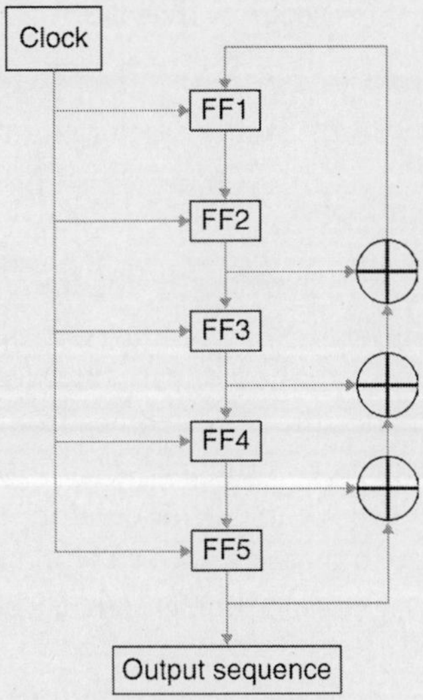

Fig. 5.3: Feedback shift registers with (5, 4, 3, 2) connection

5.2.2.1 Advantage of Selecting the Feedback Taps

1. Because of selecting the feedback taps as per Table 5.1, the PN sequence satisfies all the properties of maximum length sequences.
2. The feedback logic becomes very easy for implementation.

5.3 DSSS WITH COHERENT BPSK

In long-distance or satellite-based DSSS transmission, BPSK modulation is a suitable modulation technique that can be used.

5.3.1 BPSK Transmitter

The transmitter of DSSS with BPSK is shown in Fig. 5.4. In the transmitter, the binary data are given as an input to the non-return to zero (NRZ) level encoder. It is a coding technique in which the bit 1 is represented positively charged voltage and the bit 0 is represented by negatively charged voltage. Because the signal never goes back to zero in the midst of the bit, it is referred to as NRZ. This encoder converts the input data sequence into NRZ waveform.

The sequence generated by the PN sequence generator is encoded into the NRZ signal. The multiplier helps in doing this by multiplying the PN sequence and the coded signal. The multiplier's output is the message signal (direct sequence spread signal). This signal is provided to the BPSK modulator as a modulating signal. At the output, the direct sequence BPSK (or DS/BPSK) signal is formed.

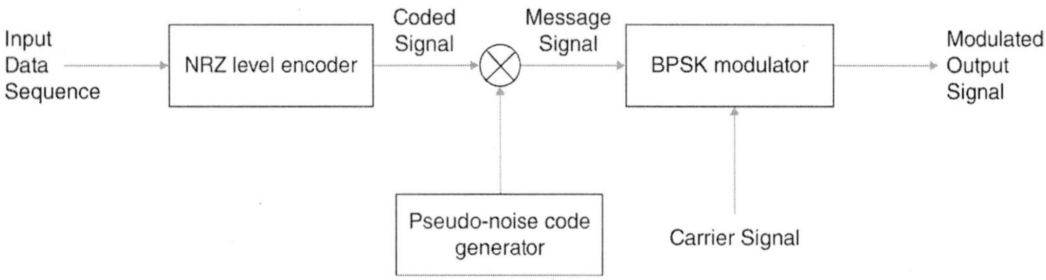

Fig. 5.4: DSSS BPSK transmitter or an encoder

5.3.2 DSSS BPSK Receiver

The DS/BPSK receiver is shown in Fig. 5.5. The block is divided into two stages. The received signal is given as input to the multiplier and multiplied with the coherent carrier. The carrier in the transmitter is synchronized with the carrier in the receiver. The low pass filter is used to filter only the low frequency signal (signal carrying message signal) after the signal comes from the multiplier. This low pass filter's cutoff frequency is the same as the frequency of the message transmission. This stage is known as the BPSK detector. This signal is fed into the second demodulator, which despreads it. The focused PN signal is a carbon copy of the one utilized in the transmitter. Over one bit period, the integrator combines the product of the recognized message signal and the PN signal. The decision is then made based on the polarity (sign) of the integrator's output.

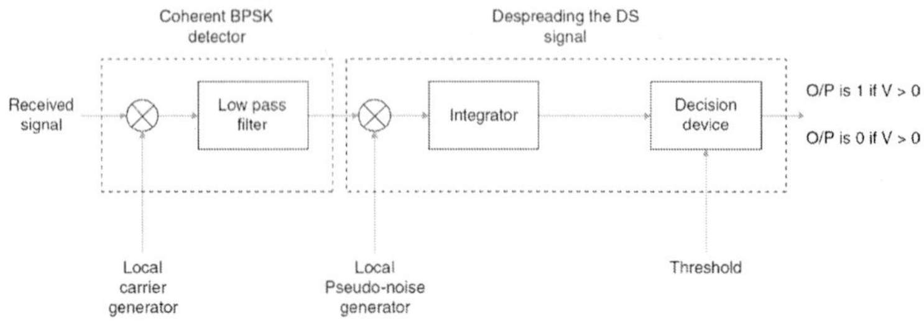

Fig. 5.5: DSSS BPSK receiver or a decoder

5.4 PERFORMANCE OF DSSS SYSTEMS

The processing gain and probability of error can be used to evaluate the performance of the DSSS system.

5.4.1 Processing Gain

Processing gain (PG) is characterized as the ratio of spread signal bandwidth to de-spreaded data signal bandwidth, i.e.

$$\text{Processing gain} = \frac{\text{BW(Spreaded signal)}}{\text{BW(Unspreaded signal)}}$$

For the NRZ signal, the bandwidth is equal to $\frac{1}{T_b}$. For the data signal (which is spreaded), bandwidth is equal to $\frac{1}{T_c}$.

The spreading PN signal is augmented by the data signal to produce the spreaded message signal. So, the processing gain is given as,

$$\text{Processing gain} = \frac{\frac{1}{T_c}}{\frac{1}{T_b}}$$

$$= \frac{T_b}{T_c}$$

The 'Gain' obtained by evaluating a spread spectrum wave on an unspread signal is known as the processing gain. We are aware that a single data bit period, T_b, represents 'N' bit periods of transmitting PN, i.e. $T_b = NT_c$.

Substituting the value T_b, we have processing gain = N (sometimes).

5.4.2 Probability of Error for the DS/BPSK System

The likelihood of error for coherent BPSK system can be determined by,

$$P(e) = \frac{1}{2} \text{ erfc}\left(\sqrt{\frac{E_b}{N_0}}\right)$$

where $\frac{N_0}{2}$ is the power spectral density of noise and E_b is the bit energy.

For DSSS modulation, the power spectral density of noise is given as,

$$\frac{N_0}{2} = \frac{JT_c}{2},$$

or $N_0 = JT_c$.

where J is the average interference power. Therefore, the probability of error becomes

$$P(e) = \frac{1}{2}\,\text{erfc.}\left(\sqrt{\frac{E_b}{JT_c}}\right)$$

5.5 FHSS SIGNALS

In the context of DSSS modulation, a broad PN sequence is multiplied with a narrower data signal. This operation results in the signal being spread across the entire available bandwidth consistently. The system's ability to resist interference is determined by a parameter called the processing gain, denoted as 'N', which is essentially the length of the PN sequence. When we increase the number of chips (N) per data bit, the output signal's bandwidth also increases, resulting in a higher processing gain. However, there are practical limitations in generating very extensive PN sequences with physical devices. Consequently, achieving extremely large bandwidths becomes challenging with DSSS modulation. To address this limitation, the frequency hop spread-spectrum technique is employed.

5.5.1 Principle of the Frequency Hop Spread Spectrum

Frequency hopping is the process of transferring data bits using different frequency slots. The total signal bandwidth is the sum of each of these frequency intervals or 'hops.' Unwanted receivers must search the whole output spectrum since the frequency hopping sequence is randomized and only identifiable by the transmitting and authorized receiving parties. Because this bandwidth encompasses the GHz range, it is difficult for accidental receivers to acquire the frequency hop data. As a result, in a FHSS, the carrier's frequency fluctuates erratically between different frequencies. MFSK is commonly used modulation in conjunction with frequency hop spread spectrum. The frequency hop spread spectrum is classified into two forms.

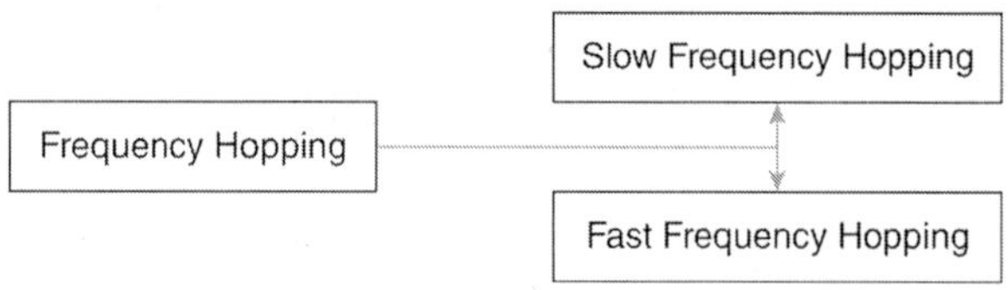

5.5.2 Slow Frequency Hopping

Slow frequency hopping, we transmit multiple symbols of data within a single frequency hop or slot. This means that the rate at which these frequency hops occur, known as the hop rate is slower than the symbol rate. Slow frequency hopping entails delivering multiple symbols in a single frequency hop, hence the name 'slow' frequency hopping.

In the context of slow frequency hopping, we start with a sequence of 'k' bits from our data input. These 'k' bits allow us to represent 'M' different symbols, where $M = 2^k$. If we want to apply spread-spectrum modulation, we take the M-ary FSK signal and further modify it to create a wider bandwidth signal. In this spread-spectrum approach, the carrier frequency randomly changes or 'hops' from one frequency to other.

Hop rate (R_h) – The hop rate is the rate at which frequency 'hops' change.

Symbol rate (R_s) – The symbol rate is the rate during which k-bit symbols in the input data sequence are created.

5.5.3 Transmitter of FH/MFSK

In Fig. 5.6, we have a block diagram illustrating a transmitter for FH-SS MFSK. Here is how it works:

We begin with a data in binary form sequence as input, which is then processed by an M-ary FSK modulator. Based on the input symbol, this modulator generates a specific frequency (one of '*M*' possible frequencies). The FSK modulator's output is then delivered to a mixer. At the same time, a precise frequency from a frequency synthesizer is delivered into the mixer. This synthesizer frequency is known as a 'hop.' When these two signals are mixed in the mixer, we receive a new signal with a frequency component equal to the total of the FSK signal and the frequency hop from the synthesizer. The FH/MFSK signal is transmitted across a wideband channel.

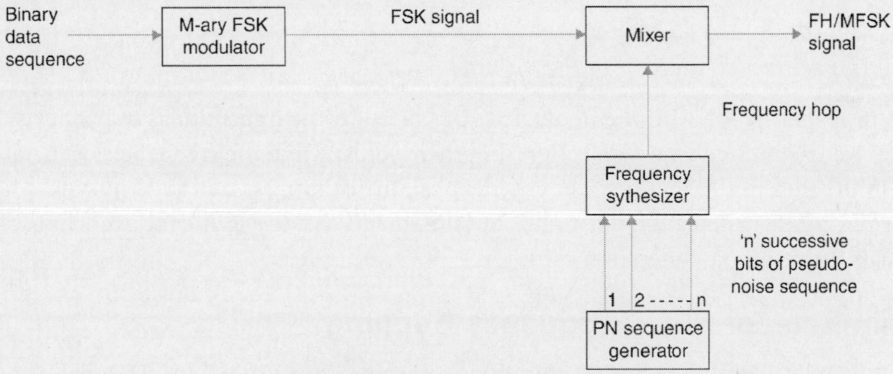

Fig. 5.6: FH/MFSK transmitter

Now, the frequency synthesizer creates the frequency hops (or slots) that we provide to the mixer. A PN sequence generator provides the inputs that operate the frequency synthesizer. The frequency hops produced by the synthesizer are determined by a series of '*n*' consecutive bits produced by this PN sequence generator. The frequency hops also fluctuate randomly because the bits that come from the PN sequence generator do. We generate a total of '2^n' unique frequency hops since these '*n*' bits govern the frequency hops. This indicates that the output signal's overall bandwidth is equal to the total of these individual frequency jumps. The aggregate bandwidth of these frequency hop signals is very large and is in the Giga Hertz (GHz) region.

5.5.4 Receiver of FH/MFSK

The FH/MFSK signal is taken in the receiver as shown in Fig. 5.7 and sent to a mixer. We simultaneously feed the frequency synthesizer's output to the same mixer. We obtain sum frequencies and difference frequencies when these signals are combined in the mixer. However, we limit the frequencies that can exit the mixer to those that are different. The information we wish to detect is carried by the M-ary FSK signals, which are these differential frequencies. The non-coherent M-ary FSK detector is then used to determine which symbol was communicated by sending these signals to it.

5.5.4.1 Reason for Non-coherent Detection

The local PN sequence generator in the receiver produces the same PN sequence as the transmitter did, but it does not produce frequencies which are perfectly synchronized

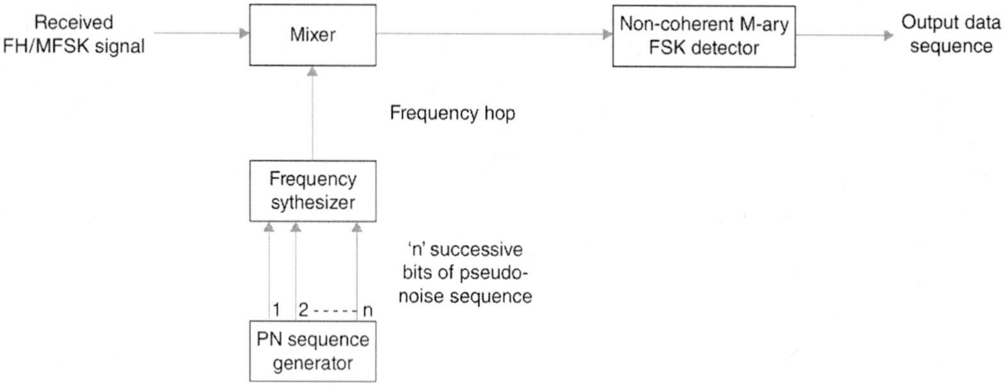

Fig. 5.7: FH/MFSK receiver

with respect to of their phase with the frequencies produced by the transmitter's frequency synthesizer, which is why we use non-coherent detection. In other words, while the frequencies generated by the synthesizer in the transmitter and receiver are identical, their timing is not quite in sync. The receiver cannot carry out coherent detection due to this misalignment. As a result, the receiver chooses to detect the FSK signals through non-coherent means.

5.5.5 Chip Rate for Slow Frequency Hopping

The specific frequency with the shortest duration is referred to as a 'chip' in FH/MFSK. Each hop in slow frequency hopping transmits numerous symbols. Each symbol will have its own unique frequency inside a single hop. For slow frequency hopping, we can state that chip rate 'R_c' and symbol rate 'R_s' are equivalent.

i.e.,
$$R_c = R_s = \frac{R_b}{k}$$

Here, R_b is the input bits rate and 'k' number of bits per symbol.

Processing gain: If N is the number of frequency hops available and K is the number of symbols transmitted in each hop, then processing gain (PG) $= \frac{N}{K}$. An example of slow frequency hopping is given below. Assume 3 bits of input binary data represent one symbol. There are eight different symbols present. If the number of hops (N) is 4, then two symbols are transmitted in each frequency hop. The processing gain is equal to 2.

5.5.6 Fast Frequency Hopping

Fast frequency hopping, in simple terms, refers to a situation where multiple frequency hops occur to transmit a single symbol. This means the rate at which these frequency changes happen, known as the hop rate (R_h), is faster than the rate at which symbols are generated (R_s). The benefit of fast hopping is that it switches the carrier frequency so rapidly that it makes it challenging for any potential interference or jamming attempts to catch and disrupt the reception of a complete symbol.

5.5.6.1 Chip Rate for Fast Hopping

Since, the hop rate is higher, the chip rate is equal to the hop rate.

i.e., $$R = R_h$$

5.5.7 CDMA Systems Based on FHSS

In this section, we explore a CDMA system that utilizes FHSS signals, which is different from the earlier CDMA system using orthogonal spread spectrum (OS-SS). In this FHSS CDMA system, each transmitter–receiver pair follows a unique and pseudo-random frequency-hopping pattern. This uniqueness allows multiple users to share the communication channel simultaneously without interfering with each other.

An interesting fact is that all users in this CDMA FHSS system have similar equipment, including encoders, decoders, modulators, and demodulators. This makes it suitable for mobile users because it is not significantly affected by timing errors, which can be problematic in OS-SS systems.

Moreover, CDMA FHSS systems can hop over a broad bandwidth, providing a high processing gain compared to OS-SS systems. This means FHSS CDMA systems offer several advantages over OS-SS CDMA systems, including increased capacity and the ability to transmit more information at higher rates.

5.5.8 Comparison Between Slow and Fast FH

A comparison between slow frequency hopping and fast frequency hopping is given in the following table.

Slow frequency hopping	Fast frequency hopping
One frequency hop transmits several symbols.	It takes several hops to send a single symbol.
Chip rate is equivalent to the symbol rate.	Chip rate is equivalent to the hop rate.
Symbol rate is higher than the hop rate.	Hop rate exceeds the symbol rate.
Over the same carrier frequency, one or more symbols are transmitted.	A single symbol is sent via several carriers in several hops.
If the carrier frequency for one hop is known, a jammer can pick up this signal.	Since one symbol is sent on numerous carrier frequencies, this signal is challenging to identify.

5.6 APPLICATIONS OF SPREAD-SPECTRUM MODULATION

In this section, we explore various applications of spread-spectrum modulation beyond its military use:

(i) Jamming resistance – Spread-spectrum modulation is known for its ability to withstand deliberate jamming or interference. While this was initially designed for military purposes, it is now also employed in commercial applications that benefit from this anti-jamming capability.

(ii) Low probability of intercept (LPI) – In military contexts, spread-spectrum is used to maintain a low signal spectral density, making it challenging for adversaries to easily detect the presence of the signal. This LPI feature enhances the security of military communications.

(iii) Mobile communications – Spread spectrum finds application in mobile communication systems because it effectively counters the adverse effects of multipath fading. Due to its large bandwidth, only a little part of the signal experiences fading, making it a valuable consideration for reliable mobile communication.

(iv) Security – Spread-spectrum communications are inherently secure. They use PN codes to generate signals, making it difficult for unintended recipients to recognize or intercept the spread-spectrum transmissions. This security feature is valuable in both military and commercial applications.

(v) Distance measurement – Spread-spectrum signals are utilized in distance measurement applications. These signals, with their broad bandwidth, enable more precise time-based measurements. Consequently, they are used in radar and navigation systems for accurate distance measurement.

(vi) Selective calling – In selective calling settings, when a central station communicates with numerous receiving points, spread spectrum is used. This is accomplished using PN codes embedded in spread-spectrum signals.

5.7 CODE DIVISION MULTIPLE ACCESS

Spread spectrum plays a pivotal role in CDMA communication systems. In CDMA, multiple users can communicate simultaneously over the same channel. Each user is assigned a specific code, and these codes are used to separate and distinguish the signals at the receiver. CDMA offers the advantage that messages from other users remain private, allowing numerous users to communicate concurrently over a shared channel.

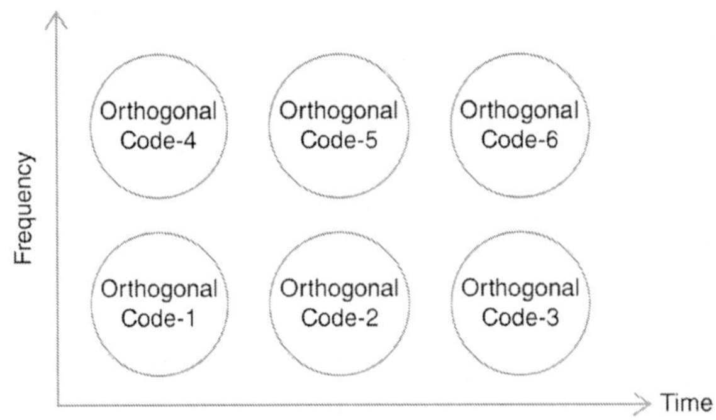

Fig. 5.8: Principle of the CDMA

Each user in this technique receives a different code pattern or signature sequence. With the use of this code, the signal is subsequently dispersed across the entire frequency range. With the aid of the same code, the signal can be recovered at the receiver. CDMA is also known as spread-spectrum multiple access (SSMA) since its signals span the whole frequency band. The user is randomly given access. As a result, frequency and time of signal transmissions from different sources overlap. The CDMA principle is shown in Fig. 5.8.

Advantages

1. The channel is used to the fullest extent possible.
2. There is no requirement for synchronization.

Disadvantages

1. Probability of data collision due to overlap.
2. In order to prevent collision, protocols are required.

EXERCISES

Two mark questions

1. List out a few applications of the spread spectrum.
2. What is the difference between an averaging and avoidance system?
3. List out the features of a spread-spectrum system.
4. What are the advantages of selecting the feedback taps?
5. What is the difference between slow frequency hopping and fast frequency hopping?
6. What are the advantages and disadvantages of using CDMA?
7. What is the full form of LPI and what is its significance?
8. List out the types of a spread-spectrum system, which is classified based on the modulation techniques used.
9. What is run property of maximum length sequence?
10. What is an NRZ level encoder?

Five mark questions

1. Explain the working of direct sequence spread spectrum with coherent BPSK transmitter and receiver.
2. Explain the receiver of FH/MFSK and the reason for using non-coherent detection.
3. List out the applications of spread-spectrum modulation and explain them briefly.
4. Explain maximum length sequences and its properties in detail.

Numerical Problems

Two mark questions

1. What is the period (in digits) of the maximum length sequence (N) if the number of shift registers used is 3?
2. What is the time period of maximum length sequence if the chip rate is 8 chips/s and the number of shift registers is 4?
3. In a DSSS communication system, the spreading code has a chip rate of 10 Mbps, and the data signal has a bit rate of 1 Mbps. Calculate the processing gain of this DSSS system.
4. In a slow frequency hopping FH/MFSK system, the input bit rate (R_b) is 1 Mbps, and each symbol consists of 4 bits ($k = 4$). Calculate the chip rate (R_c) and symbol rate (R_s) for this system.

KNOW MORE

- The FHSS method is used by a number of ultra-low power (ULP) radio protocols to get past interference problems in the crowded 2.4 GHz band (Nordic devices, for instance, use FHSS-based Bluetooth low energy and nRF24L Series protocols). To carry out the process, the 2.4 GHz band, which in reality ranges from 2.4 to 2.48 GHz, is separated into channels with different frequencies.
- When interference is detected, both the transmitter and the receiver switch to a new and presumably clearer frequency. The transmitter transmits on a certain channel. FHSS is easy to use, economical and efficient. It may be assumed to be contemporary, but FHSS actually predates much of modern wireless technology by around a century.
- Hedy Lamarr was a much sought-after leading lady in Hollywood throughout the 1940s. Outside of the lens of the cameras, however, her spirit of invention gave rise to the wireless communication technologies we take for granted today. Hedwig Kiesler, better known as Hedy Lamarr, was born in Vienna, in 1914, and immigrated to the United States, in 1937.
- On set, Hedy Lamarr would devote hours testing hypotheses and playing with technology in her trailer. One of these concepts would fundamentally alter wireless communication, laying the groundwork for technologies like Wi-Fi, Bluetooth, and, of course, GPS.
- 'Frequency hopping' was a clever technique for changing between radio frequencies to prevent signal jamming. It was created as a 'secret telecommunications system' by Hedy Lamarr and American composer George Antheil.

REFERENCES AND SUGGESTED READINGS

1. Simon Haykin: Digital Communications, John Wiley & Sons, New Delhi, 2017

References for further reading

1. https://onlinecourses.nptel.ac.in/noc20 ee34/preview

2. https://onlinecourses.nptel.ac.in/noc20_ee34/preview

Lab Experiments

List of Experiments

Through this module, the following experiments are designed and developed: The objective of each experiment is

- to modulate a high-frequency carrier with a sinusoidal signal to obtain an amplitude modulated (AM) signal, calculate the modulation index, observe the spectrum and calculate the bandwidth.
- to modulate a high-frequency carrier with a sinusoidal signal to obtain a frequency modulated (FM) signal, calculate the modulation index, observe the spectrum and calculate the bandwidth.
- to study and observe the operation of a super heterodyne receiver.
- to modulate a pulse carrier with a sinusoidal signal to obtain pulse width modulated (PWM) signal and demodulate it.
- to modulate a pulse carrier with a sinusoidal signal to obtain pulse position modulated signal (PPM) and demodulate it.
- verification of the sampling theorem, PPM techniques, (flat top and natural sampling), reconstruction of original signal, observe aliasing effect in a the frequency domain.
- to observe the operation of a pulse code modulation (PCM) encoder and decoder. Find the theoretical and simulated values of SQNR.
- to study and observe the amplitude response of an automatic gain controller (AGC).

I. A Review of Signals and Transforms Using MATLAB

Objectives:

- Generation and study of various sequences such as unit impulse, unit step and waveforms (both periodic and aperiodic) such as square, saw tooth, triangular, sinusoidal, ramp, and sinc.

Software/Equipment Required: MATLAB

Theory:

1. **Definition of a signal:**
    - A signal is a physical quantity that varies with time, space, or any other independent variable. It can represent information or data.
1. **Continuous-Time Signals (Analog):**
    - **Mathematical Representation:**
        - ➢ $x(t)$ represents a continuous-time signal, where t is a real-valued variable representing time.

- **Amplitude:** The values of $x(t)$ represent the amplitude of the signal at each instant.
- **Example:**
  - A sine wave $x(t)=A\sin(2\pi ft+\phi)$, where A is the amplitude, f is the frequency, t is the time and ϕ is the phase.

2. **Discrete-Time Signals (Digital)**
   - **Mathematical Representation**
     - $x[n]$ represents a discrete-time signal, where n is an integer representing discrete instances of time.
   - **Sequence:** $x[0], x[1], x[2], \ldots, x[n], \ldots$
   - **Example**
     - A unit step sequence $u[n]$ is 1 for $n \geq 0$ and 0 for $n < 0$.
3. **Common Signal Operations**
   - **Amplitude Scaling**
     - $y(t) = c\,x(t)$ (for continuous-time signals)
     - $y[n] = c\,x[n]$ (for discrete-time signals)
   - **Time Scaling (Compression/Expansion)**
     - $y(t) = x(at)$ (for continuous-time signals)
     - $y[n] = x(an)$ (for discrete-time signals)
   - **Time Shifting**
     - $y(t) = x(t - t_0)$ (for continuous-time signals)
     - $y[n] = x[n - n_0]$ (for discrete-time signals)
   - **Time Reversal**
     - $y(t) = x(-t)$ (for continuous-time signals)
     - $y[n] = x[-n]$ (for discrete-time signals)
4. **Signal Classifications**
   - **Deterministic Signals**
     - Can be precisely predicted
     - Example: sinusoidal signals
   - **Random (Stochastic) Signals**
     - Have an element of unpredictability
     - Example: noise in a communication channel
5. **Periodic and Aperiodic Signals:**
   - **Periodic Signals:**
     - $x(t) = x(t + T)$ for all t, where T is the period.
   - **Aperiodic Signals**
     - Do not exhibit a repetitive pattern.
6. **Impulse and Step Functions:**
   - **Unit Impulse Function $\delta(t)$ or $\delta[n]$:**
     - $\delta(t) = 0$ for $t \neq 0$, $\int_{-\infty}^{\infty} \delta(t)\,dt = 1$.

- **Unit Step Function $u(t)$ or $u[n]$:**
  - $u(t) = 0$ for $t < 0$, $u(t) = 1$ for $t \geq 0$.

CODE

```
% Generation of signals and sequences
clc;
clear all;
close all;
%~~~~~~~~~~~~~~~~~~~~~~~~~~~~~~~~~~~~~~~
%generation of unit impulse signal
t1=-1:0.1:1;
y1=(t1==0);
subplot(1,2,1);
plot(t1,y1,'linewidth',1);
xlabel('time');
ylabel('amplitude');
title('unit impulse signal');
%generation of impulse sequence
subplot(1,2,2);
stem(t1,y1,'linewidth',1);
xlabel('n');
ylabel('amplitude');
title('unit impulse sequence');
```

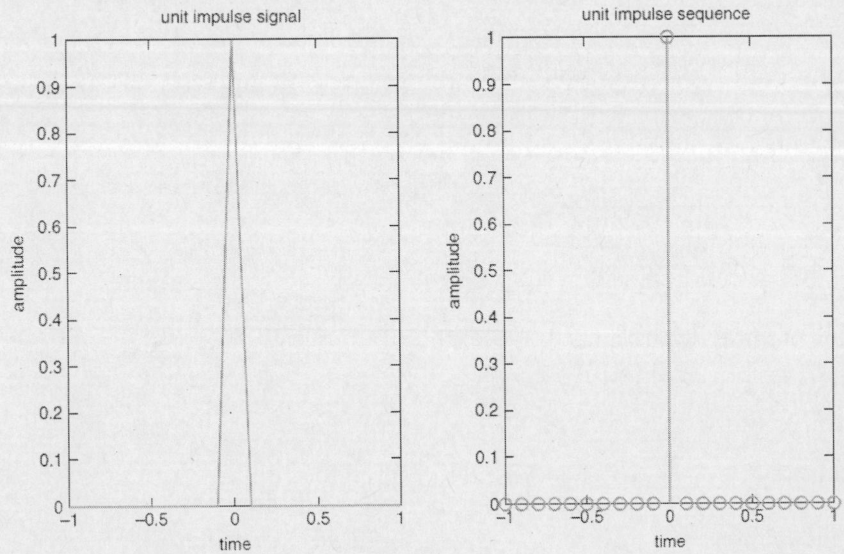

```
%~~~~~~~~~~~~~~~~~~~~~~~~~~~~~~~~~~~~~~~~~~~~~
%generation of unit step signal
t2=-10:1:10;
y2=(t2>=0);
subplot(1,2,1);
plot(t2,y2, 'linewidth', 1);
xlabel('time');
ylabel('amplitude');
title('unit step signal');
%generation of unit step sequence
subplot(1,2,2);
stem(t2,y2,'linewidth', 1);
xlabel('n');
ylabel('amplitude');
title('unit step sequence');
```

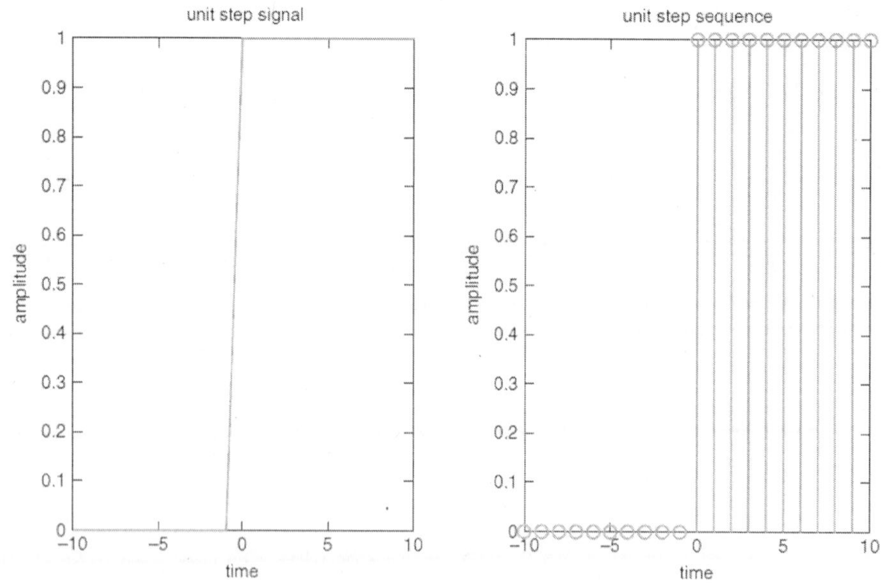


```
%~~~~~~~~~~~~~~~~~~~~~~~~~~~~~~~~~~~~~~~~~~~~~
%generation of square wave signal
t=0:0.001:0.1;
y3=square(2*pi*20*t);
figure;
subplot(1,2,1);
plot(t,y3,'linewidth',1);
```

```
axis([0 0.1 -1.5 1.5]);
xlabel('time');
ylabel('amplitude');
title('square wave signal');
%generation of square wave sequence
subplot(1,2,2);
stem(t,y3,'linewidth',1);
axis([0 0.1 -2 2]);
xlabel('n');
ylabel('amplitude');
title('square wave sequence');
```

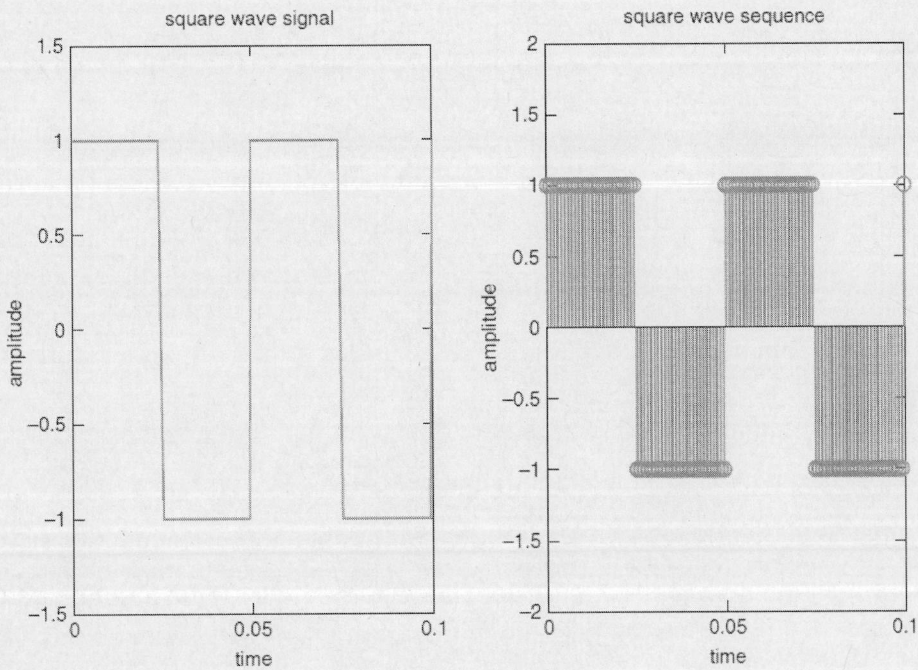


```
%~~~~~~~~~~~~~~~~~~~~~~~~~~~~~~~~~~~~~~~~~~
%generation of sawtooth signal
t=0:0.002:0.1;
y4=sawtooth(2*pi*50*t);
subplot(1,2,1);
plot(t,y4,'linewidth',1);
axis([0 0.1 -1 1]);
```

```
xlabel('time');
ylabel('amplitude');
title('sawtooth wave signal');
%generation of sawtooth sequence
subplot(1,2,2);
stem(t,y4,'linewidth',1);
axis([0 0.1 -1  1]);
xlabel('n');
ylabel('amplitude');
title('sawtooth wave sequence');
```

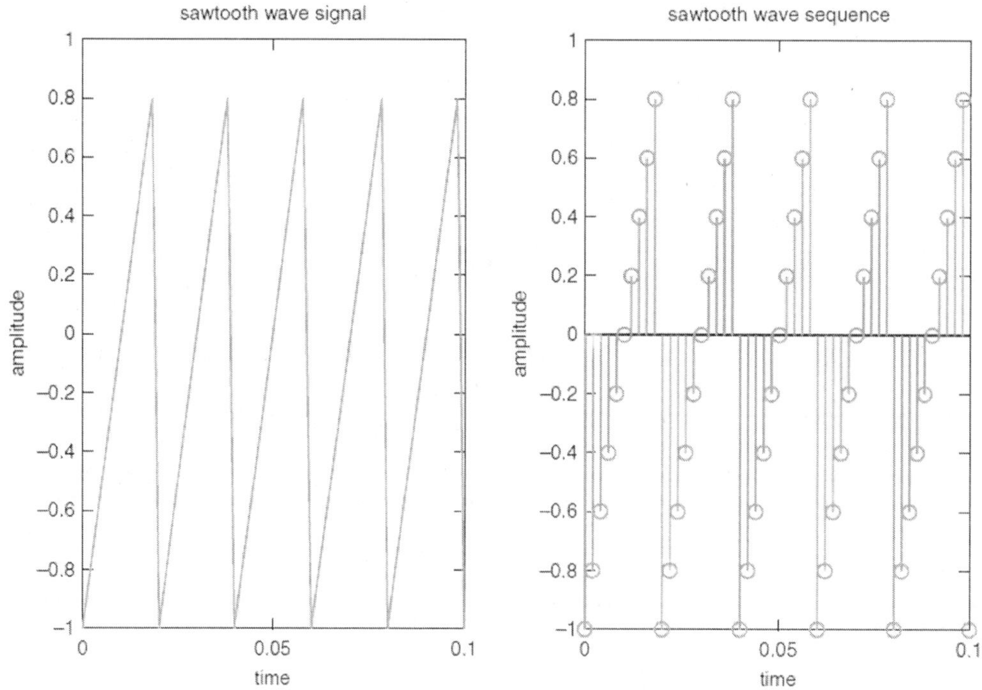


```
%~~~~~~~~~~~~~~~~~~~~~~~~~~~~~~~~~~~~~~~~~~~
%generation of a triangular wave signal
t=0:0.002:0.1;
y5=sawtooth(2*pi*50*t,.5);
figure;
subplot(1,2,1);
plot(t,y5,'linewidth',1);
axis([0 0.1 -1.5 1.5]);
```

```
xlabel('time');
ylabel('amplitude');
title(' triangular wave signal');
%generation of a triangular wave sequence
subplot(1,2,2);
stem(t,y5,'linewidth',1);
axis([0 0.1 -1.5 1.5]);
xlabel('n');
ylabel('amplitude');
title('triangular wave sequence');
```

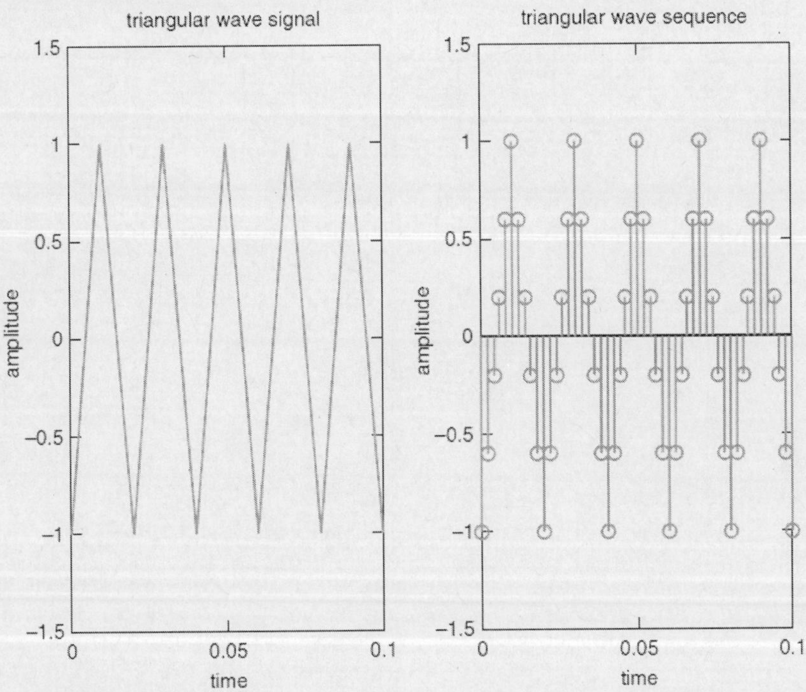


```
%~~~~~~~~~~~~~~~~~~~~~~~~~~~~~~~~~~~~~~~~~~
%generation of sinusoidal wave signal
t=0:0.002:0.1;
y6=sin(2*pi*40*t);
subplot(1,2,1);
plot(t,y6,'linewidth',1);
axis([0 0.1 -1.5 1.5]);
xlabel('time');
```

```
ylabel('amplitude');
title(' sinusoidal wave signal');
%generation of sin wave sequence
subplot(1,2,2);
stem(t,y6,'linewidth',1);
axis([0 0.1 -1.5 1.5]);
xlabel('n');
ylabel('amplitude');
title('sine wave sequence')
```

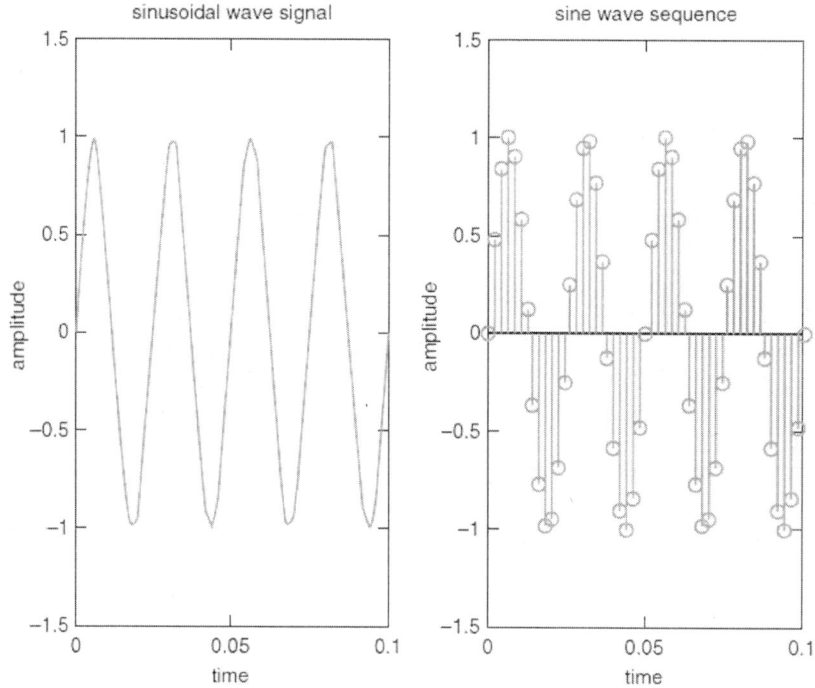


```
%~~~~~~~~~~~~~~~~~~~~~~~~~~~~~~~~~~~~~~~~~~~
%generation of ramp signal
t=0:0.002:0.1;
y7=t;
figure;
subplot(1,2,1);
plot(t,y7,'linewidth',1);
xlabel('time');
ylabel('amplitude');
title('ramp signal');
```

```
%generation of ramp sequence
subplot(1,2,2);
stem(t,y7,'linewidth',1);
xlabel('n');
ylabel('amplitude');
title('ramp sequence');
```

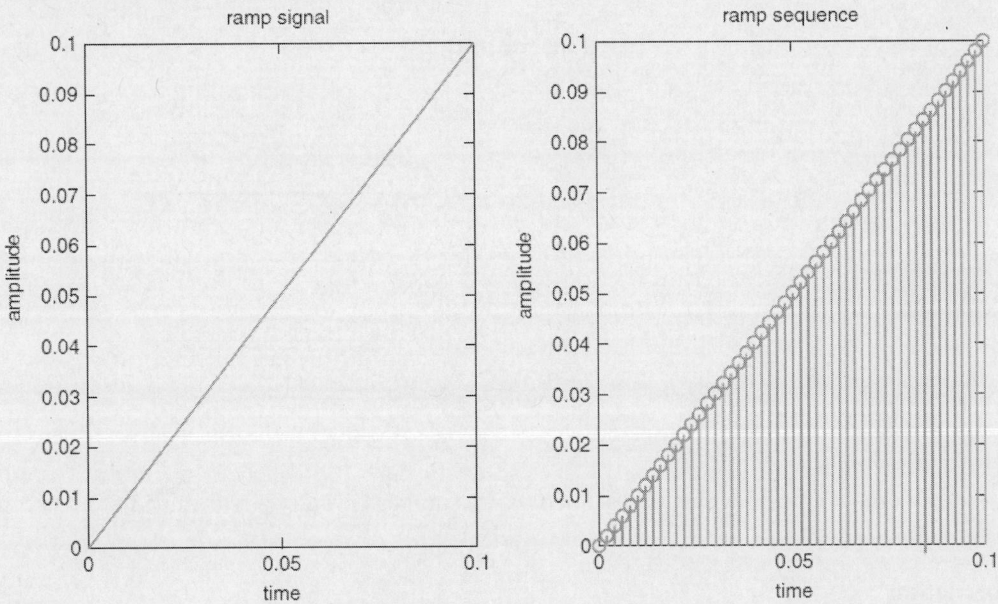


```
%~~~~~~~~~~~~~~~~~~~~~~~~~~~~~~~~~~~~~~~~~~
%generation of sinc signal
t=0:0.002:0.1;
t3=linespace(-5,5);
y8=sinc(t3);
subplot(1,2,1);
plot(t3,y8,'linewidth',1);
xlabel('time');
ylabel('amplitude');
title(' sinc signal');
%generation of sinc sequence
subplot(1,2,2);
stem(y8,'linewidth',1);
xlabel('n');
ylabel('amplitude');
title('sinc sequence');
```

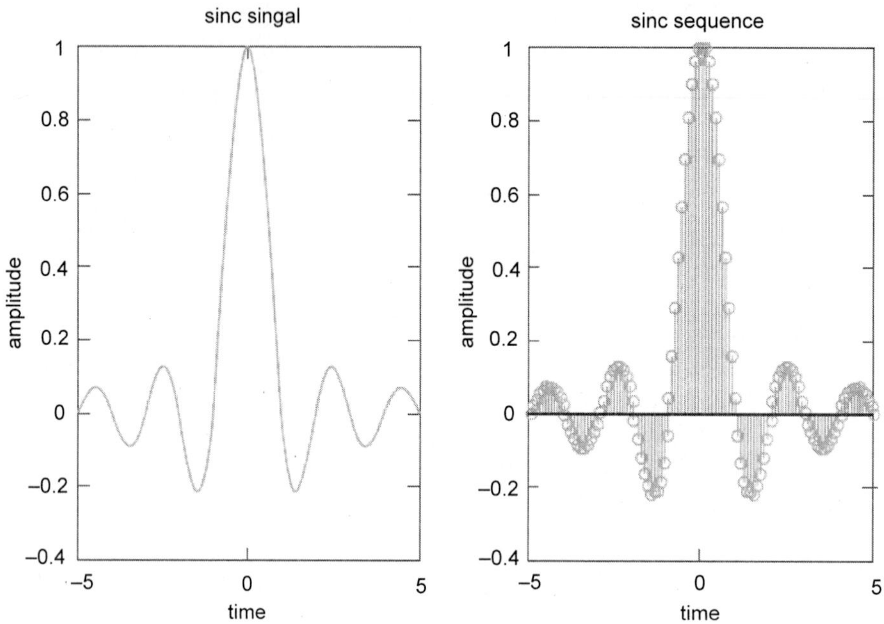

Correlations of sequences

It is a measure of the degree to which two sequences are similar. Given two real-valued sequences $x(n)$ and $y(n)$ of finite energy.

Auto-correlation

$$r_{x,x}(l) = \sum_{n=-\infty}^{+\infty} x(n)x(n-l)$$

where l is the shift or lag parameter.

Cross-correlation

$$r_{x,y}(l) = \sum_{n=-\infty}^{+\infty} x(n)y(n-l)$$

CODE

```
clc;
clear all;
close all;
% two input sequences
x=input('enter input sequence');
h=input('enter impulse sequence');
subplot(2,2,1);
stem(x);
```

```
xlabel('n');
ylabel('x(n)');
title('input sequence');
subplot(2,2,2);
stem(h);
xlabel('n');
ylabel('h(n)');
title('impulse sequence');
% cross-correlation between two sequences
y=xcorr(x,h);
subplot(2,2,3);
stem(y);
xlabel('n');
ylabel('y(n)');
title(' cross-correlation between two sequences ');
% auto-correlation of the input sequence
z=xcorr(x,x);
subplot(2,2,4);
stem(z);
xlabel('n');
ylabel('z(n)');
title('auto-correlation of the input sequence');
% cross-correlation between two signals
% generating two input signals
t=0:0.2:10;
x1=3*exp(-2*t);
h1=exp(t);
figure;
subplot(2,2,1);
plot(t,x1);
xlabel('t');
ylabel('x1(t)');
title('input signal');
subplot(2,2,2);
plot(t,h1);
xlabel('t');
ylabel('h1(t)');
```

```
title('impulse signal');
% cross-correlation
subplot(2,2,3);
z1=xcorr(x1,h1);
plot(z1);
xlabel('t');
ylabel('z1(t)');
title('cross-correlation ');
% auto-correlation
subplot(2,2,4);
z2=xcorr(x1,x1);
plot(z2);
xlabel('t');
ylabel('z2(t)');
title('auto-correlation ');
clc;
clear all;
close all;
% two input sequences
x=input('enter input sequence');
h=input('enter impulse sequence');
subplot(2,2,1);
stem(x);
xlabel('n');
ylabel('x(n)');
title('input sequence');
subplot(2,2,2);
stem(h);
xlabel('n');
ylabel('h(n)');
title('impulse sequence');
% cross- correlation between two sequences
y=xcorr(x,h);
subplot(2,2,3);
stem(y);
```

```
xlabel('n');
ylabel('y(n)');
title('cross-correlation between two sequences');
% auto- correlation of the input sequence
z=xcorr(x,x);
subplot(2,2,4);
stem(z);
xlabel('n');
ylabel('z(n)');
title('auto-correlation of the input sequence');
% cross- correlation between two signals
% generating two input signals
t=0:0.2:10;
x1=3*exp(-2*t);
h1=exp(t);
figure;
subplot(2,2,1);
plot(t,x1);
xlabel('t');
ylabel('x_1(t)');
title('input signal');
subplot(2,2,2);
plot(t,h1);
xlabel('t');
ylabel('h_1(t)');
title('impulse signal');
% cross-correlation
subplot(2,2,3);
z1=xcorr(x1,h1);
plot(z1);
xlabel('t');
ylabel('z_1(t)');
title('cross-correlation ');
% auto-correlation
subplot(2,2,4);
z2=xcorr(x1,x1);
plot(z2);
```

```
xlabel('t');
ylabel('z_2(t)');
title('auto-correlation');
```

OUTPUT

enter the input sequence [1 2 3 4]

enter the impulse sequence [5 6 7 8 9]

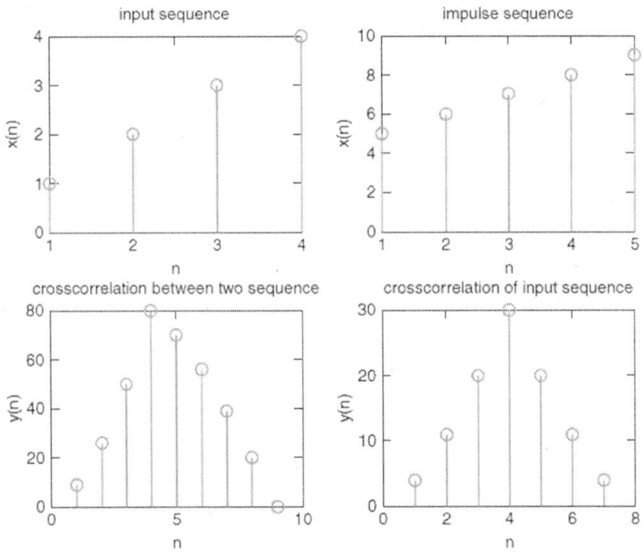

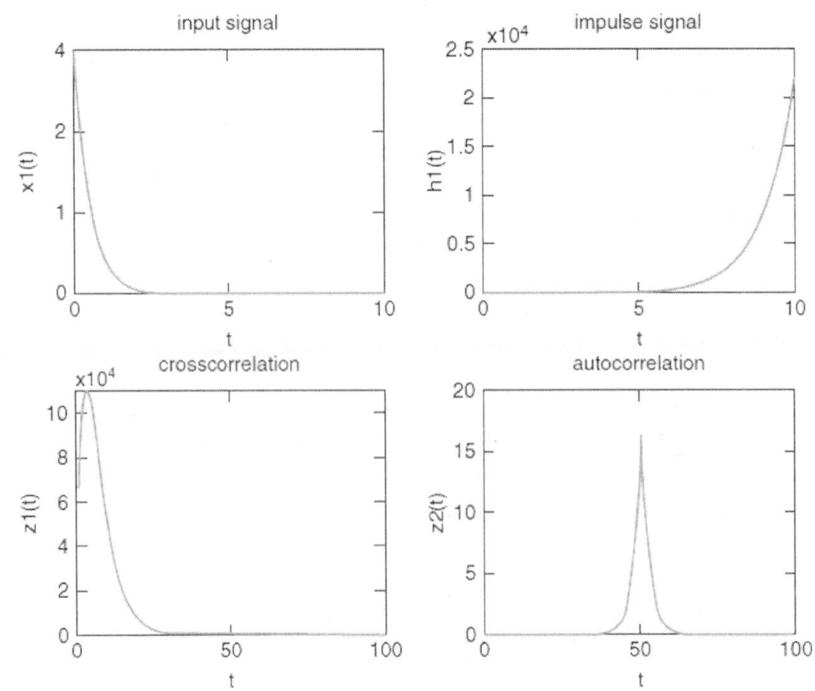

Fourier Transform

The Fourier Transform is a mathematical tool used in signal processing and mathematics to analyse the functions and signals in the frequency domain. It decomposes a function or a signal into its constituent frequencies. The transform is named after Joseph Fourier, who first introduced the concept in the early 19th century. The Fourier transform is widely used in various fields, including signal processing, image analysis, communication systems, audio processing, and many other areas where understanding the frequency content of a signal is important.

The Fourier transform of a continuous function $f(t)$ is defined as:

$$F(\omega) = \int_{-\infty}^{+\infty} f(t) e^{-j\omega t} dt$$

Here, $F(\omega)$ is the Fourier transform of $f(t)$, ω is the angular frequency, and j is the imaginary unit. The inverse Fourier Transform allows us to go from the frequency domain back to the time domain.

CODE

```
clc;
clear all;
close all;
fs=1000;
N=1024; % length of the FFT sequence
t=[0:N-1]*(1/fs);
% input signal
x=0.8*cos(2*pi*100*t);
subplot(3,1,1);
plot(t,x);
axis([0 0.05 -1 1]);
grid;
xlabel('t');
ylabel('amplitude');
title('input signal');
% Fourier transform of given signal
x1=fft(x);
% magnitude spectrum
k=0:N-1;
Xmag=abs(x1);
subplot(3,1,2);
```

```
plot(k,Xmag);
grid;
xlabel('t');
ylabel('amplitude');
title('magnitude of FFT signal')
%phase spectrum
Xphase=angle(x1);
subplot(3,1,3);
plot(k,Xphase);
grid;
xlabel('t');
ylabel('angle');
title('phase of FFT signal');
```

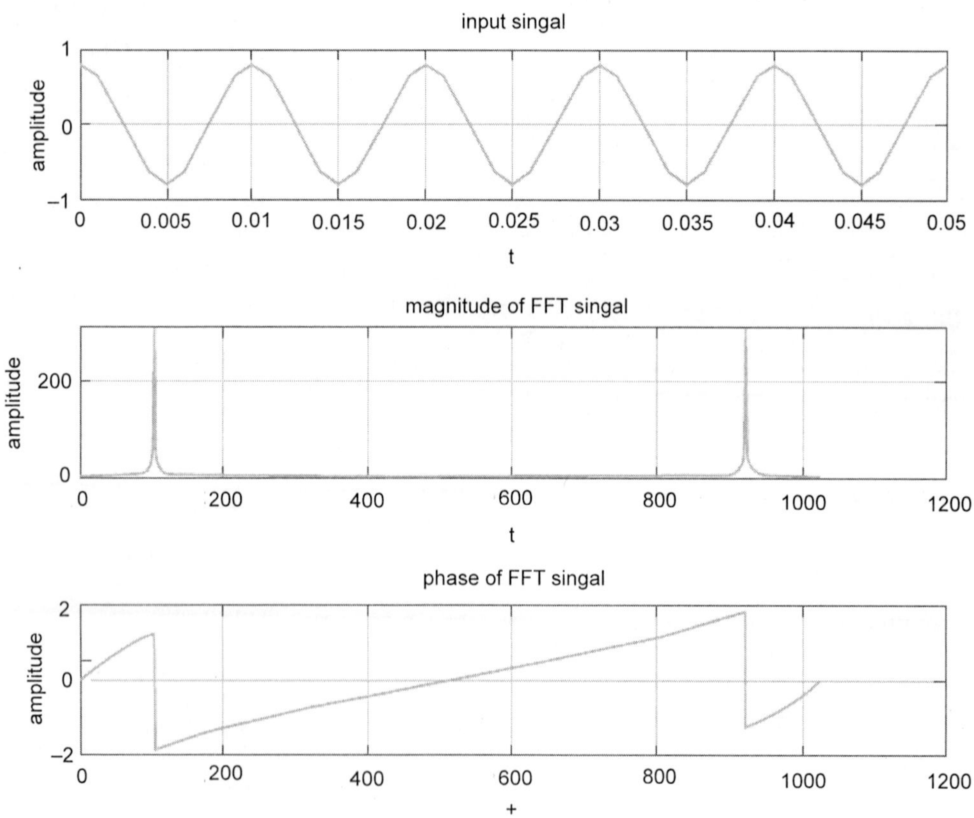

Result

Thus, in this experiment, all types of signals and sequences, auto-correlation and cross-correlation and Fourier transform operation are studied.

Amplitude Modulation and Demodulation

Objectives

- Using a transistor-based mixer circuit, generate an AM waveform.
- Calculate the modulation index for the resulting waveform.
- Using an envelope detector, retrieve the original message signal.

Components/Equipment Required

S. No.	Components	Range	Quantity
1.	Function generator	0–15 MHz	2
2.	Digital signal oscilloscope (DSO)	0–60 MHz	1
3.	Diode	OA 79	1
4.	PNP transistor	AC 128	1
5.	Inductor	5 mH	1
6.	Resistor	5.6 kΩ	1
7.	Capacitor	0.01 μF	1
		0.01 μF	2
8.	Oscilloscope probes	–	3

Design

The carrier frequency can be found using the following equation:

$$f_c = 1/(2\pi\sqrt{LC})$$

Theory

A communication system aims to exchange information from source to the destination. The information-carrying signal is shifted to higher frequency before transmission, which is called as modulation. The carrier signal is responsible for the long distance transmission. The choice of the carrier signal depends upon the application.

Message Signal

It is the raw data such as voice, image or video signal and usually of lower frequency. A simple message signal is denoted by

$$m(t) = A_m \sin(2\pi f_m t)$$

where A_m is the amplitude of the message signal and f_m is the frequency of the message signal.

Carrier Signal

It assists in carrying or transmitting the message signal to its destination without any information loss. Carrier signals are typically high-frequency sinusoidal signals.

$$c(t) = A_c \sin(2\pi f_c t)$$

where A_c is the amplitude of the carrier signal, and f_c is the frequency of the carrier signal.

Amplitude Modulation

The change in magnitude of the carrier signal varies according to the amplitude of the message signal at any given time. The amplitude of the carrier signal is used for AM.

At the receiver, the envelope of the message signal can be obtained from the envelope of the modulated signal. Note that the message, carrier, and modulated signals are all analog signals.

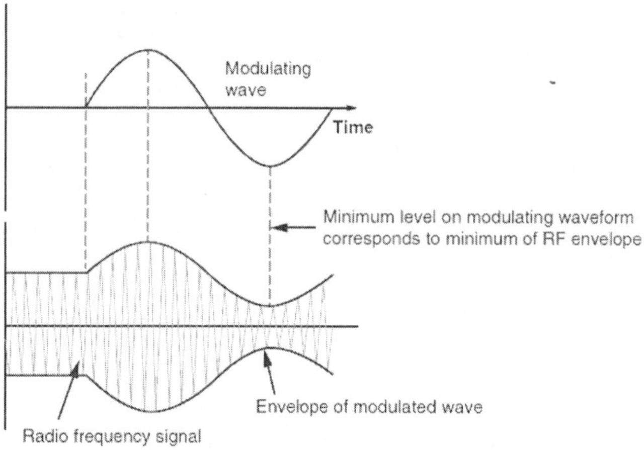

Amplitude Modulation

Modulation Index

It is a property of the modulated signal. It describes the measure of movement of a modulated variable about its unmodulated level. In AM scheme, it is given by

AM index, —

From the diagram, it is clear that A_{max} and A_{min} are the maximum and minimum amplitudes of the modulates signal, respectively, which are expressed as

$A_{max} = A_m + A_c$, $A_{min} = A_c - A_m$.

So, the modulation index is $\mu = \dfrac{A_{max} - A_{min}}{A_{max} + A_{min}}$.

Based on the value of μ, modulation is classified into three types,

1. If $\mu < 1$, it is said to be under modulation.
2. If $\mu = 1$, it is said to be critical modulation.
3. If $\mu > 1$, it is said to be over-modulation.

Characteristics of modulated signal under different types of modulated signal

- For under-modulated signal,

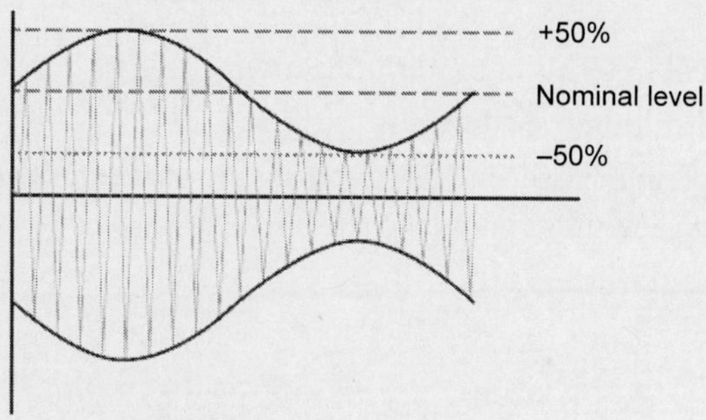

Amplitude modulated index of 0.5

Here, the carrier signal is not completely utilized for modulation.

- For critical modulated signals,

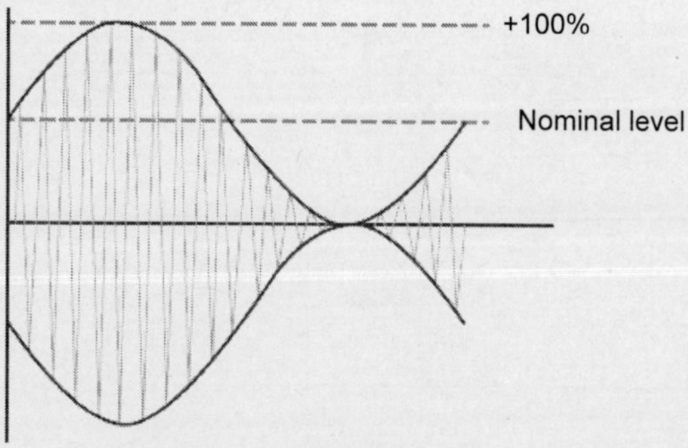

Amplitude modulated index of 1.0

Here, the carrier is utilized to 100% for modulation.

- For over-modulated signals,

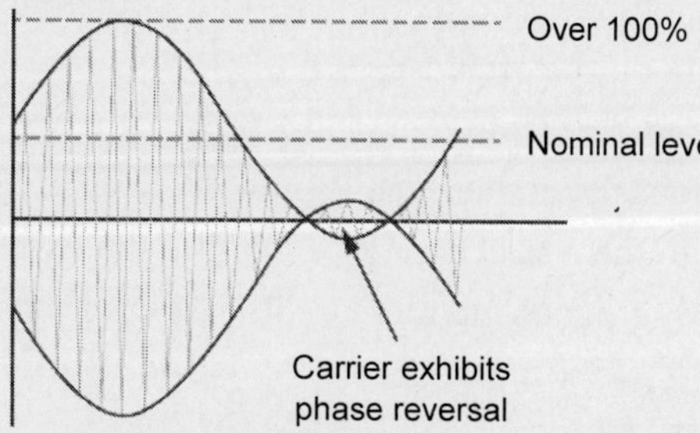

Amplitude modulated index of more than 1.0

Here, the carrier suffers phase reversal and adopts negative values, as well as the introduction of some additional sidebands, that result in an expansion of bandwidth and interference for users.

Advantages of Amplitude Modulation

- AM is more straightforward to implement.
- AM receivers are less expensive because fewer specialized components are required, and the circuit has few components.

Disadvantages of AM

- AM could be more efficient in terms of power utilization.
- AM needs to use its bandwidth more efficiently.
- Because most noise is amplitude-based, and AM receivers are more sensitive to amplitude, AM signals are more susceptible to high amounts of noise.

Circuit Diagram

AM generator circuit

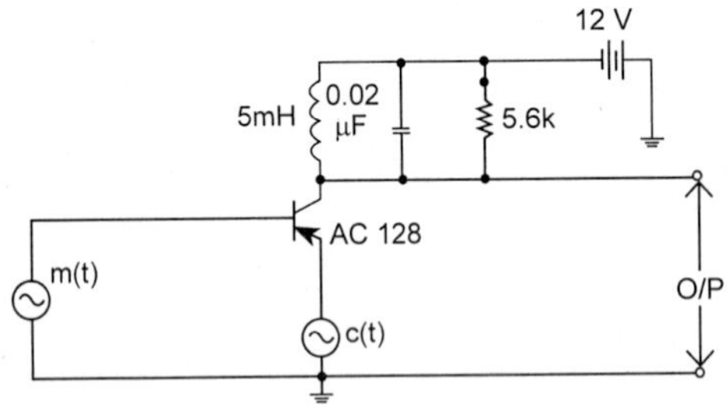

Envelope detector

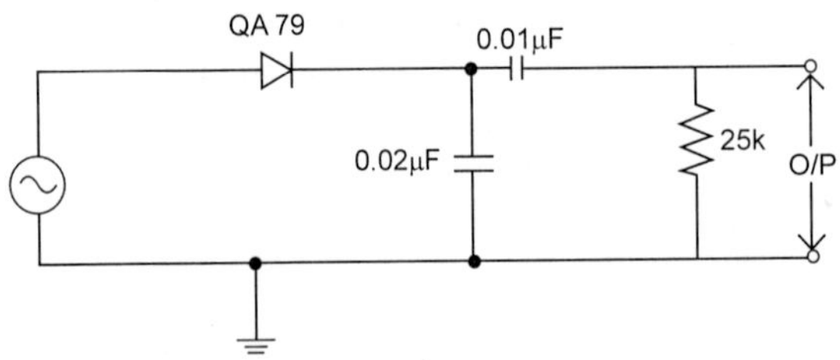

Procedure

1. Construct the AM generator circuit.
2. Set appropriate values for amplitudes and frequencies of modulating and carrier waves.
3. Observe the output waveform on the oscilloscope.
4. Observe the values of A_{max} and A_{min} from the oscilloscope.
5. Calculate the modulation index using the given formula.
6. Apply the output waveform to the AM generation circuit to the envelope detector circuit.
7. Observe the output modulating signal and measure its amplitude and frequency.

Observation

Signal	Amplitude (V)	Frequency (Hz)
Modulating signal		
Carrier signal		
AM signal		
Reconstructed modulating signal		

Results

Thus, the amplitude modulation and demodulation circuit are designed and developed. The parameters are

1. Modulation index =
2. Amplitude of the demodulated signal =
3. Frequency of the demodulated signal =

Frequency Modulation and Demodulation

Objective

- To simulate the FM signal and demodulate it
- To determine the modulation index of FM
- To verify the Carson's rule from the frequency spectrum.

Software Required

MATLAB 2023

Theory

Frequency Modulation

It is a type of angle modulation where the frequency of the carrier wave changes in response to the magnitude of the message signal. FM is an analog modulation technique that is popularized utilized.

Single-tone Frequency Modulation

Let $m(t)$ be the modulating signal with magnitude V_m and frequency f_m,

$$m(t) = V_m \cos(2\pi f_m t)$$

The carrier signal is given by,

$$c(t) = V_c \cos(2\pi f_c t + \varnothing)$$

In FM, depending on the change in voltage of the modulating wave, the frequency of the carrier changes. Therefore, the instantaneous frequency is given by,

$$f_i(t) = f_c + k_f\, m(t)$$

$$f_i(t) = f_c + k_f V_m \cos(2\pi f_m t)$$

$$f_i(t) = f_c + \Delta f \cos(2\pi f_m t)$$

where $\Delta f = k_f V_m$ is the frequency deviation.

The FM index is given by

$$\beta_f = \frac{\Delta f}{B}$$

where B is the bandwidth of the message signal.

Carson's Rule

A Wideband FM has infinite bandwidth theoretically. This is not possible practically. Hence, to find the bandwidth of FM signal, Carson's rule is used.

$$B_T = 2B(1 + \beta_f)$$

Block Diagram

Frequency Modulation

Getting FM signal from the PM signal is given by,

$$m_f(t) = \frac{D_p}{D_f}\left(\frac{dm_p(t)}{dt}\right)$$

where D_p and D_f are the PM and FM indices, respectively.

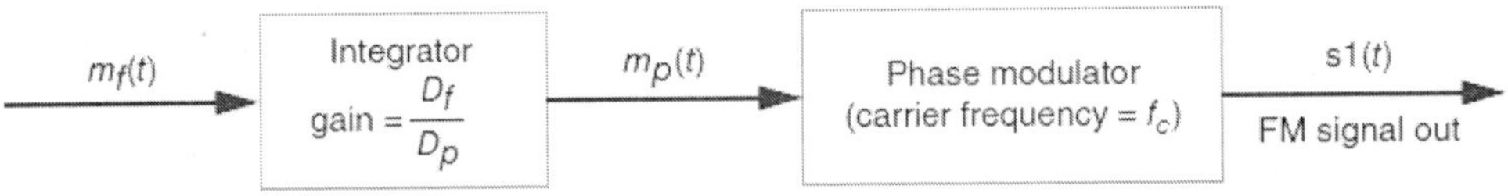

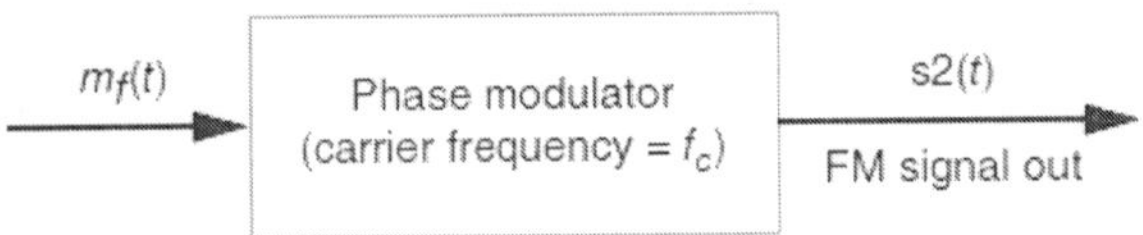

CODE

```
clear all;
clc;
msg_amp = 1;
msg_freq = 100;
carrier_freq = 5000;
% defining sampling frequency
samp_freq=(2^12)*carrier_freq;
% defining time interval
time_int=0:1/samp_freq:1/msg_freq;
% length of the time interval
```

```
N_time_int=length(time_int);
% defining frequency axis
freq_axis=0:samp_freq/N_time_int:(samp_freq-(samp_freq/N_time_int));
% defining frequency deviation constant
freq_dev=600*pi;
% defining phase sensitivity parameter
phase_sen=1;
% defining bandwidth of the message signal
B=msg_freq;
% defining maximum frequency deviation
del_f=(freq_dev*msg_amp)/(2*pi);
disp('Maximum frequency deviation');
disp(del_f);
% defining frequency modulation index
Bf=del_f/B;
disp('Frequency modulation index');
disp(Bf);
% frequency message signal
msg_sig=msg_amp*cos(2*pi*msg_freq*time_int);
% carrier signal
carrier_sig=cos(2*pi*carrier_freq*time_int);
x=msg_amp*sin(2*pi*msg_freq*time_int)/(2*pi*msg_freq);
% phase message signal
mp=(freq_dev/phase_sen)*x;
T=phase_sen*mp;
% frequency modulated signal using phase modulation
Sig1=cos((2*pi*carrier_freq*time_int)+T);
% frequency modulated signal using frequency modulation
Sig2=fmmod(msg_sig,carrier_freq,samp_freq,del_f/msg_amp);
% Bandwidth of modulated signal as per Carson's rule
Bt=2*(Bf+1)*msg_freq;
disp('Bandwidth by Carsons rule');
```

```
disp(Bt);
% demodulated signal
sig3=fmdemod(Sig1,carrier_freq,samp_freq,del_f);
figure(1);
subplot(3,2,1);
plot(time_int,msg_sig);
title('Input message signal');
subplot(3,2,2);
plot(time_int,carrier_sig);
title('Modulating carrier wave');
subplot(3,2,3);
plot(time_int,T);
title('Phase modulates signal');
subplot(3,2,4);
plot(time_int,Sig2);
title('FM Signal using Frequency Modulator');
subplot(3,2,5);
plot(time_int,Sig1);
title('FM signal using Phase Modulator');
subplot(3,2,6);
plot(freq_axis,abs(fft(Sig1)));
title('Spectrum of frequency modulated signal');
figure(2);
plot(freq_axis,abs(fft(Sig1)));
title('Spectrum of frequency modulated signal');
figure(3);
plot(time_int,sig3); title('Demodulated signal');
```

OUTPUTS

Command Window

```
Maximum frequency deviation= 300
Frequency modulation index= 3
Bandwidth by Carson's rule  = 800
```

Plots

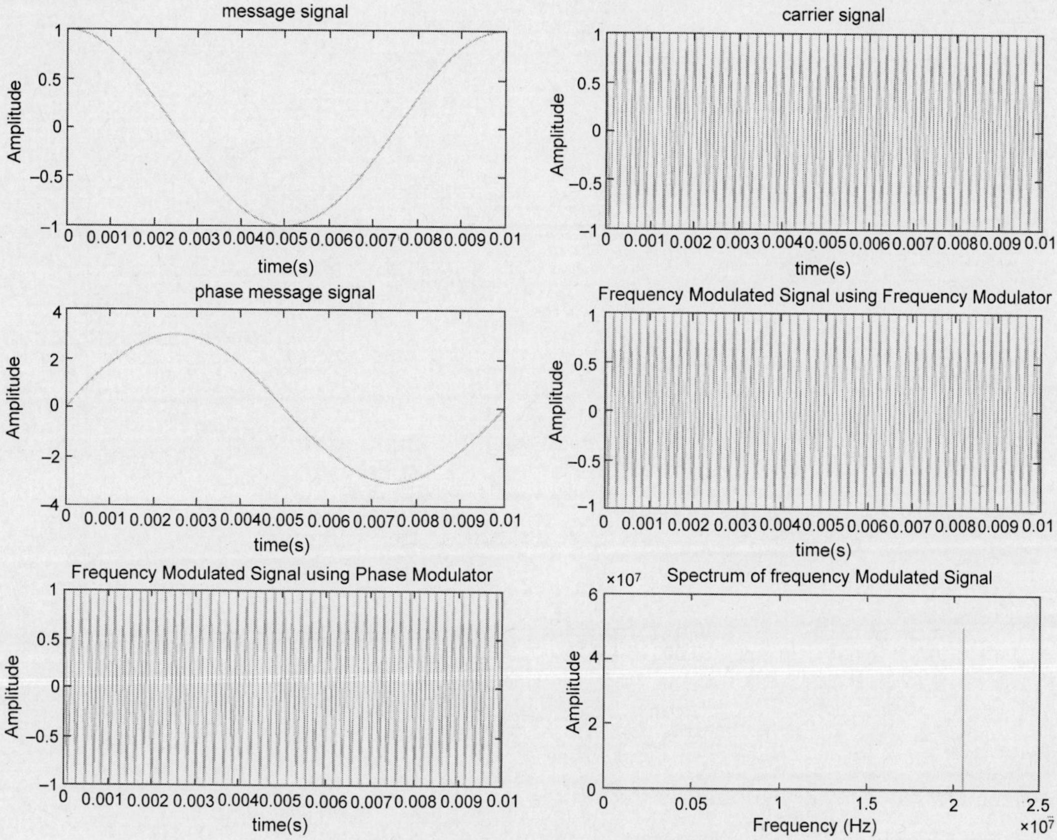
message signal
Amplitude
time(s)
carrier signal
Amplitude
time(s)
phase message signal
Amplitude
time(s)
Frequency Modulated Signal using Frequency Modulator
Amplitude
time(s)
Frequency Modulated Signal using Phase Modulator
Amplitude
time(s)
×10^7
Spectrum of frequency Modulated Signal
Amplitude
Frequency (Hz)
×10^7

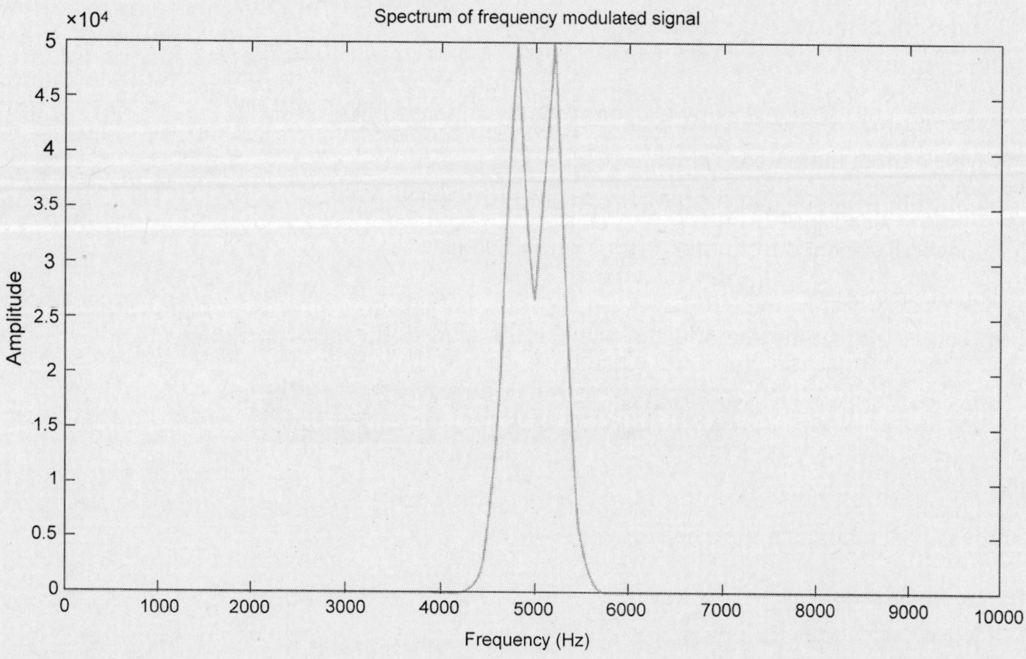
×10^4
Spectrum of frequency modulated signal
Amplitude
Frequency (Hz)

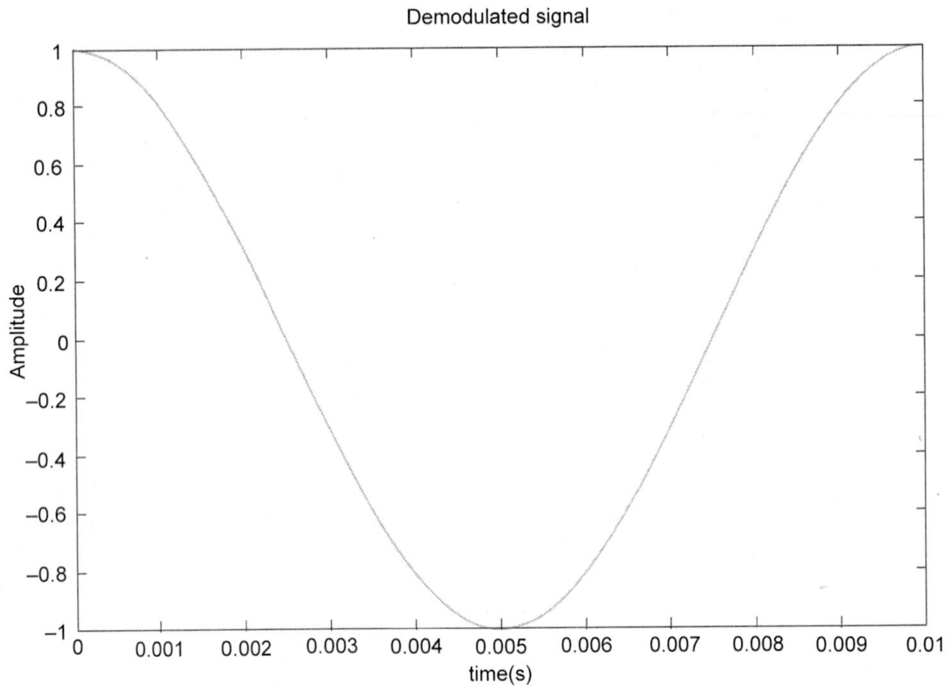

Result

- Maximum frequency deviation, $\Delta f = (600\pi/(2\pi)) = 300$ Hz
- Frequency range of FM modulated signal is from $f_c - \Delta f = 4700$ to $f_c + \Delta f = 5300$ Hz
- Bandwidth of the message signal, $B = 100$ Hz
- FM index = 3
- Bandwidth of FM signal using Carson's rule, $B_T = 2(3 + 1)100 = 800$ Hz
- Verification of the Carson's rule:
- Bandwidth from the plot = 5400 Hz – 4600 Hz = 800 Hz
- Bandwidth obtained from the Carson's rule = 800 Hz
- Thus, Carson's rule is verified.
- It is observed that the demodulated signal is the same as the message signal.

Superheterodyne Receiver

Objective

To study the operation of a superheterodyne receiver

Software Required

MATLAB 2023

Theory

Superheterodyne receiver

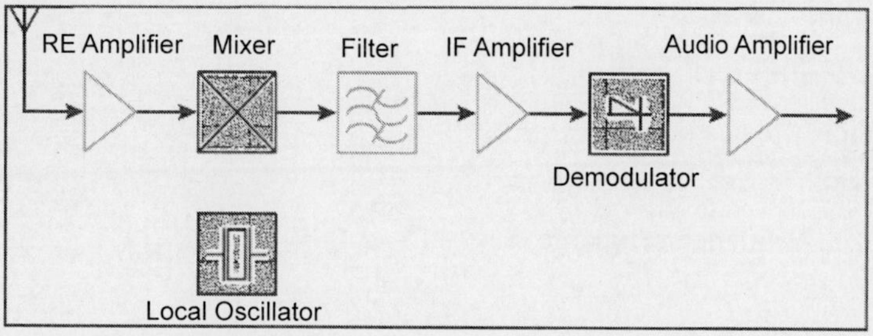

It is a receiver designed to capture radio signals and demodulate them by changing the frequency of the incoming signal to that of the center frequency of a bandpass filter. This demodulation device is composed of three distinct components:

1. **IF Mixer**

   A circuit that converts the input signal to a specific intermediate frequency.

2. **The Band-Pass Filter**

   It eliminates all frequency information except that moved to the intermediate frequency. It transmits both the required signal and the image frequency, which is a disadvantage of the receiver.

3. **Demodulator**

   The received and separated signal will be demodulated using whichever procedure is required for the modulation type of the signal.

Superheterodyne Detectors

The detector encountered fluctuations in demodulation whenever there was phase variation throughout modulation, including findings of 0 at specified moments (occurs if the carrier and incoming signal have a 90° quadrature). As a result, in prior investigations, the standard product demodulator was only used for signals that arrive with linear modulation. When modulating FM signals, the signal carrier at the receiver's end must track that of the signals being modulated. Because this is a challenging operation, a more accessible way could be to correct the filter and utilize a variable local oscillator to produce a potentially demodulated band. Combining all side harmonics into a unified constituent is required for message signal reconstruction.

CODE

```
%Constant parameters
sample_freq = 10^7;
max_time = 1/500;
n_fft_pts = 2^20;
max_freq = 150*10^3;
```

```
k = 10^3;
samp_time = 1/sample_freq;
time_repeats = 100;
%frequency axis scaling
scale_freq = sample_freq/max_freq;
if (scale_freq < 1)
   error('Your max. frequency very large')
end
%time vector creation
time_vec = 0:samp_time:time_repeats*max_time;
%Generate Signals
carrier_freq = 120*k;
het_freq = 20*k;
image_freq = 100*k;
%generate AM signal
message_sig = cos(2*pi*2*k*time_vec) + 0.333*cos(2*pi*6*k*time_vec);
carrier_sig = cos(2*pi*carrier_freq*time_vec);
am = carrier_sig .* (message_sig+1.5);
het_oscillator_sig = cos(2*pi*het_freq*time_vec);
%applying IF mixer
mixer = am .* het_oscillator_sig;
%applying band-pass filter
fil_order = 4;
bandwidth = 8*k;
fc1 = 2*(image_freq-bandwidth)/sample_freq;
fc2 = 2*(image_freq+bandwidth)/sample_freq;
[b,a] = butter(fil_order/2, [fc1, fc2]);
band_pass_f = filter(b, a, mixer);
 %applying envelope detector
fil_order = 4;
cutoff_freq = 10*k;
env = abs(band_pass_f);
```

```
[b,a] = butter(fil_order/2, 2*(cutoff_freq/sample_freq));
demod = filter(b, a, env);
%create frequency-domain vectors
freq = (0:round(n_fft_pts/scale_freq)-1);
freq = freq * (sample_freq/n_fft_pts);
x_message = abs(fft(message_sig, n_fft_pts));
x_message = x_message(1:round(n_fft_pts/scale_freq));
x_message = x_message / max(x_message);
x_am = abs(fft(am, n_fft_pts));
x_am = x_am(1:round(n_fft_pts/scale_freq));
x_am = x_am / max(x_am);
x_bpf = abs(fft(band_pass_f, n_fft_pts));
x_bpf = x_bpf(1:round(n_fft_pts/scale_freq));
x_bpf = x_bpf / max(x_bpf);
x_demod = abs(fft(demod, n_fft_pts));
x_demod = x_demod(1:round(n_fft_pts/scale_freq));
x_demod = x_demod / max(x_demod);
time_vec = time_vec(1:(size(time_vec,2)-1) / time_repeats);    %restrict vector size to max_t
message_sig = message_sig(1:(size(message_sig,2)-1) / time_repeats);
am = am(1:(size(am,2)-1) / time_repeats);
band_pass_f = band_pass_f(1:(size(band_pass_f,2)-1) / time_repeats);
demod = demod(1:(size(demod,2)-1) / time_repeats);
figure;
subplot(4,2,1)
plot(time_vec, message_sig)
title('Message Signal')
xlabel('Time (s)')
ylabel('Voltage (V)')
subplot(4,2,2)
plot(freq, x_message)
title('Message Signal')
```

```
xlabel('Frequency (Hz)')
ylabel('Magnitude')
subplot(4,2,3)
plot(time_vec, am)
title('AM-Modulated Message')
xlabel('Time (s)')
ylabel('Voltage (V)')
subplot(4,2,4)
plot(freq, x_am)
title('AM-Modulated Message')
xlabel('Frequency (Hz)')
ylabel('Magnitude')
subplot(4,2,5)
plot(time_vec, band_pass_f)
title('IF Mixing and Band-Pass Filtering')
xlabel('Time (s)')
ylabel('Voltage (V)')
subplot(4,2,6)
plot(freq, x_bpf)
title('IF Mixing and Band-Pass Filtering')
xlabel('Frequency (Hz)')
ylabel('Magnitude')
subplot(4,2,7)
plot(time_vec, demod)
title('Demodulated Message')
xlabel('Time (s)')
ylabel('Voltage (V)')
subplot(4,2,8)
plot(freq, x_demod)
title('Demodulated Message')
xlabel('Frequency (Hz)')
ylabel('Magnitude')
```

OUTPUTS

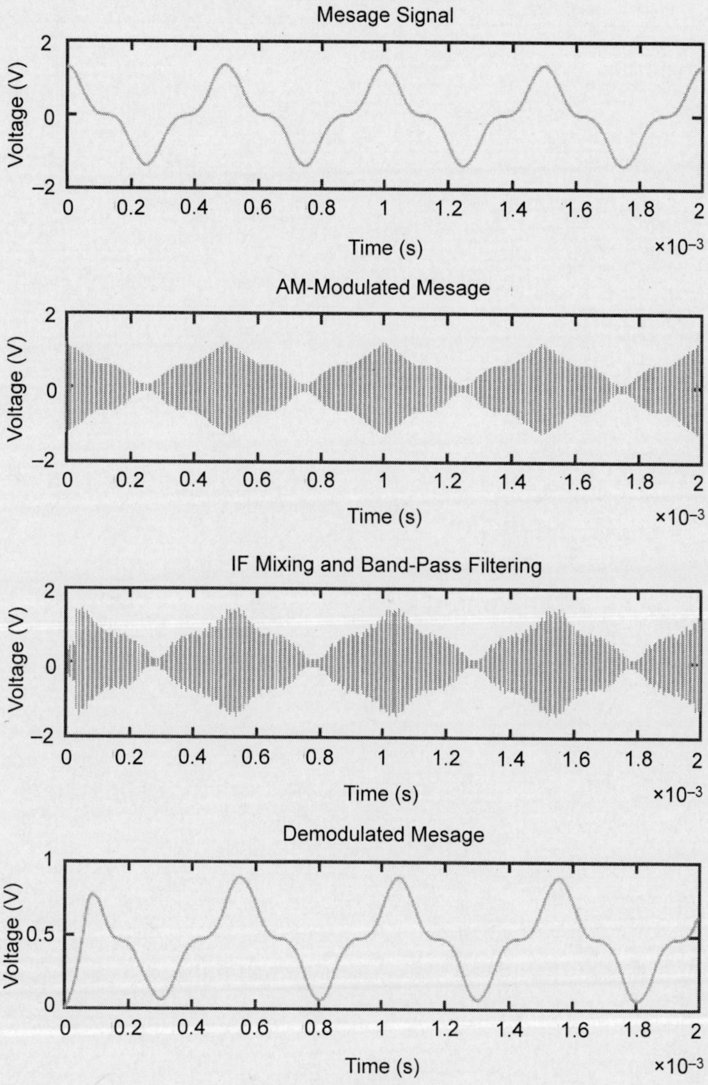
Mesage Signal
Voltage (V)
Time (s)
×10⁻³
AM-Modulated Mesage
Voltage (V)
Time (s)
×10⁻³
IF Mixing and Band-Pass Filtering
Voltage (V)
Time (s)
×10⁻³
Demodulated Mesage
Voltage (V)
Time (s)
×10⁻³

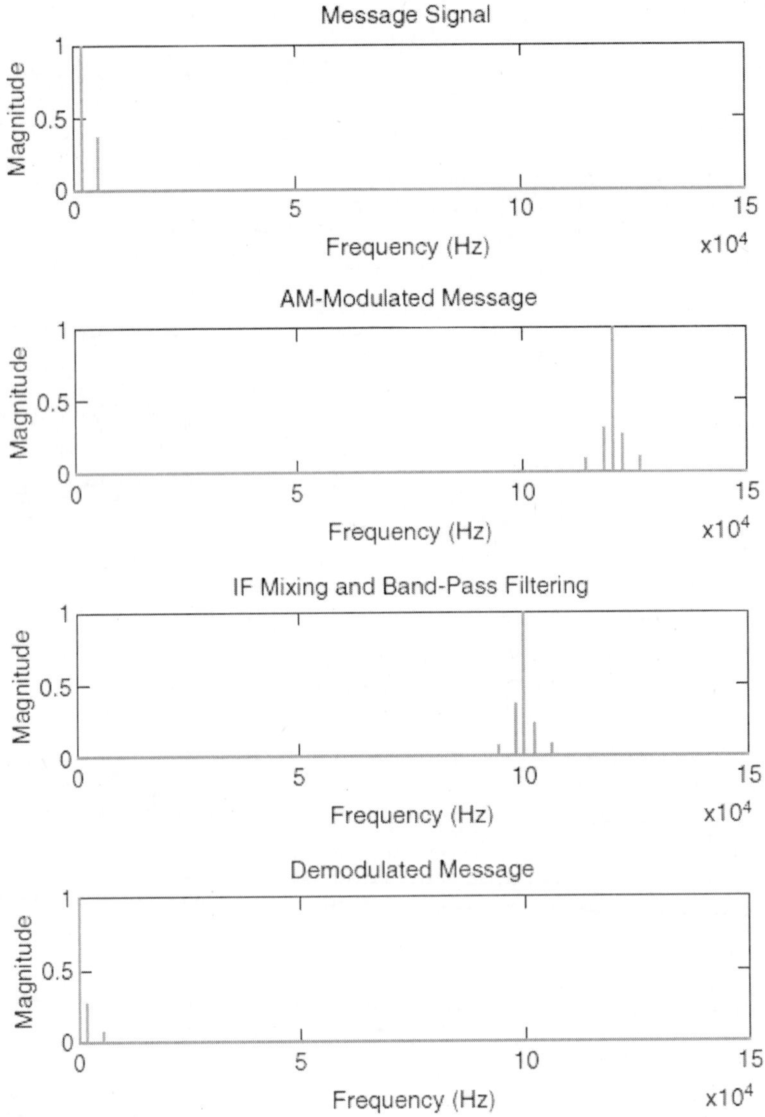

Results

The superheterodyne receiver is simulated, and time and frequency domain plots at different stages are obtained.

Pulse Width Modulation

Objective

To synthesize and demodulate pulse width modulated signals.

Components Required

S. No.	Components	Value	Quantity
1.	IC	IC555	1
2.	Resistors	1.2 kΩ	1
		1.5 kΩ	1
		8.2 kΩ	1
3.	Capacitors	1 µF	1
		0.01 µF	1
4.	Function generator	–	1
5.	Digital signal oscilloscope	–	1
6.	DC power supply	–	1

Theory

Pulse Width Modulation

Pulse width modulation (PWM) uses message signal samples to modify the duration of individual pulses. The width can be modified by altering the duration of occurrence of the leading edge, trailing edge, or both sides of the pulse in line with the modulating wave.

Circuit Diagram

Pulse Width Modulation

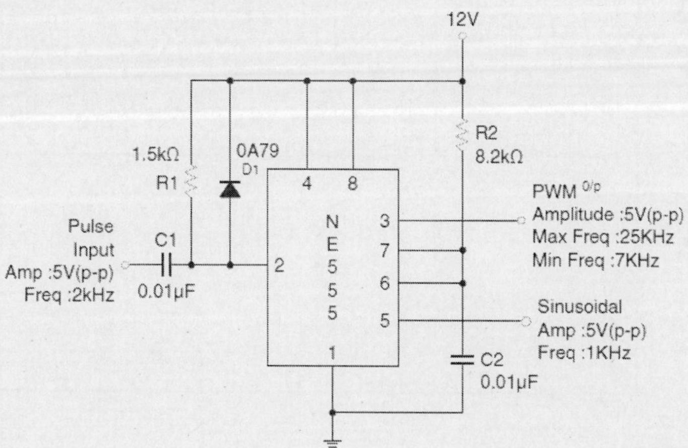

Pulse Width Demodulation

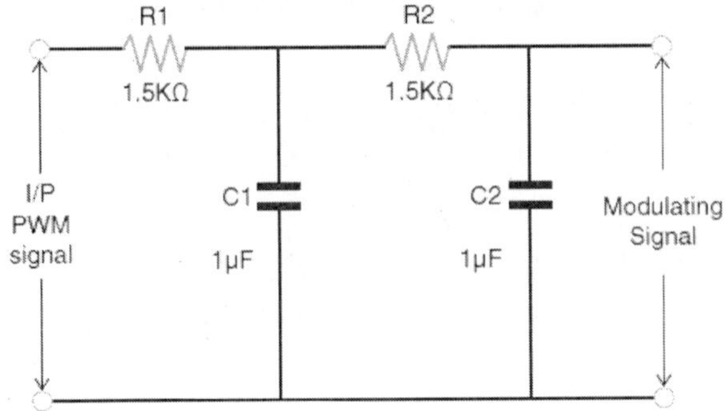

Procedure

- Fix the circuit as shown in the modulator circuit diagram.
- Generate a pulse signal with f_m = 2 kHz and A_m = 5 V (peak to peak).
- Observe the sample wave at PIN 3.
- Apply the alternating current signal to PIN 5 and change the magnitude.
- As the controlling potential changes, so does the output pulse width.
- Notice how the pulse width grows with a positive slope and decreases with a negative slope. The pulse width of a sinusoidal waveform will be most significant at the positive peak and lowest at the negative peak. Make a note of your findings.
- Apply a PWM wave to the demodulation circuit and see the demodulated wave that results.

Model Waveforms

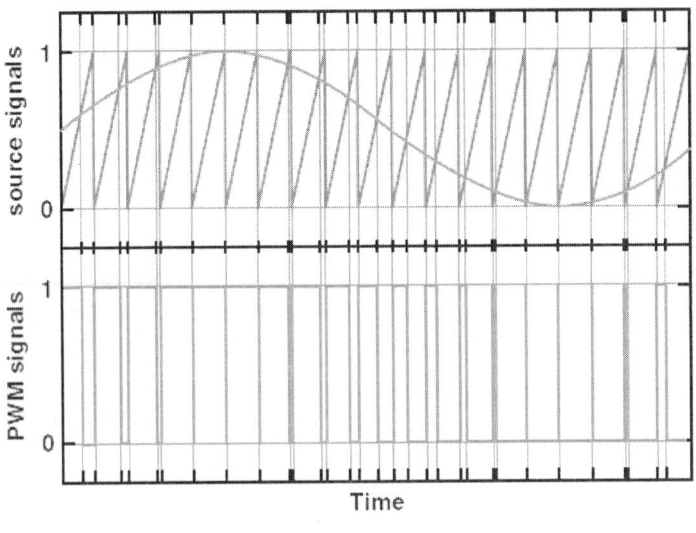

Result

The pulse width modulated wave has been generated, and demodulation is performed on the same wave.

Pulse Position Modulation

Objective

To synthesize and demodulate PPM signals.

Components/Apparatus Required

S. No.	Components	Range/Part No.
1.	IC	IC555
2.	Transistor	BC107
3.	Resistor	3 kΩ
		3.9 kΩ
		5.6 kΩ
		10 kΩ
4.	Capacitors	60 μF
		0.01 μF
5.	Function generator	–
6.	DSO	–

Theory

Pulse Position Modulation

The amplitude and duration of the pulse do not vary in PPM, but the position of the pulse varies proportionally according to the sampled values of the data. It is a signaling technique that encodes the value of an analog signal's sample data on the digital signal's temporal axis, similar to angle modulation approaches. PWM and PPM are the two significant kinds of pulse time modulation (PTM). The analog sample value determines the position of a narrow pulse relative to the clocking time in PPM. The channel bandwidth in PPM is determined by the pulse rise time. It has a minimal level of noise interference.

Circuit Diagram

Pulse Position Modulation

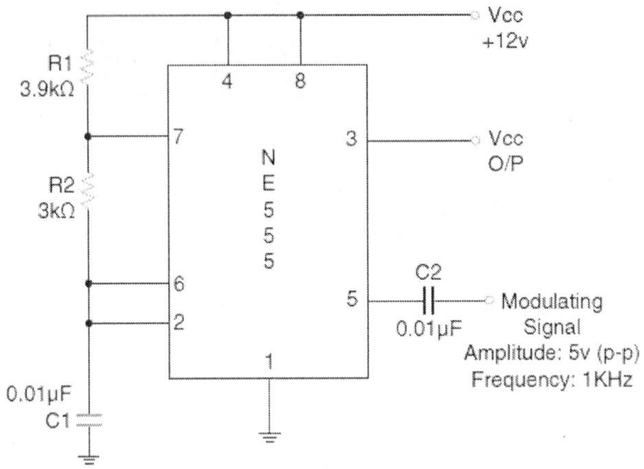

Pulse Position Signal Demodulation

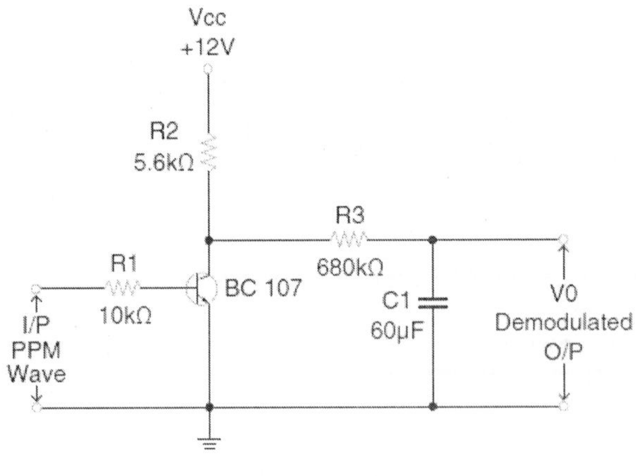

Procedure

1. Connect the circuit according to the PPM schematic.
2. Examine the sample result at PIN 3 and the location of the pulses in a digital signal oscilloscope.
3. Tune the magnitude by raising the power supply slightly. Take note of the frequency of the pulse result as well.
4. Apply the sinusoidal voltage of 2 V (peak–peak AC signal) to PIN 5 using a function generator.
5. Now, record the position of the pulses by altering the magnitude of the modulating signal.
6. Provide the PPM signal as feed to the demodulation circuit during the demodulation procedure.
7. Examine the output on a cathode ray oscilloscope.

Model Waveform

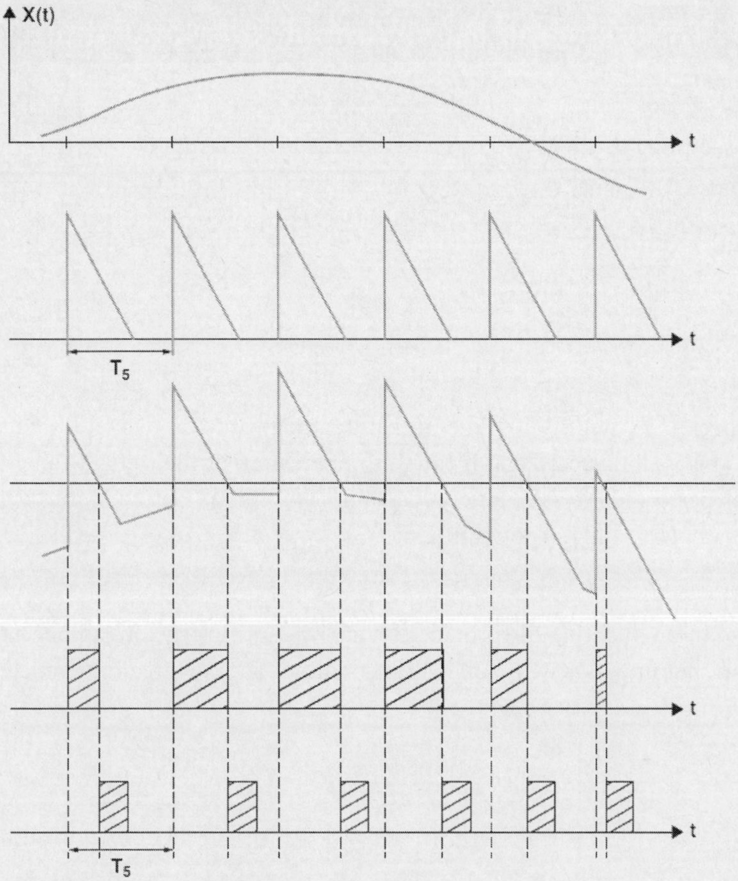

Observation

Modulating signal Magnitude (V_{p-p})	Pulse width ON (ms)	Pulse width OFF (ms)	Total time period (ms)

Result

PPM wave is generated, and demodulation is done on the same wave.

Verification of the Sampling Theorem and Pulse Amplitude Modulation

Objective

- To verify the sampling theorem and observe the effects of aliasing
- To simulate pulse AM signal.

Software Required

MATLAB 2023

Theory

Sampling

Sampling converts a continuous-time signal (analog) into a discrete-time signal with little information loss. This is done because it is tough to process all of the signal's values throughout time. Hence, sampling is performed to facilitate processing. Sampling is classified into three types: impulse sampling, natural sampling, and flat-top sampling.

Impulse Sampling

It is performed by augmenting the provided signal by an infinitely long impulse train. In principle, this sampling is fine, but impulses with null length could not be generated in practice.

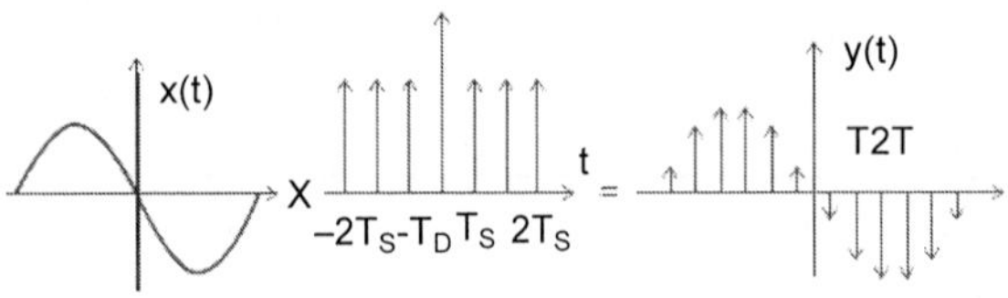

Natural Sampling

It is accomplished by multiplying the provided signal, $x(t)$, by an infinitely long pulse train. Because the magnitude of one pulse changes along a given signal, it becomes difficult to eliminate noise during propagation.

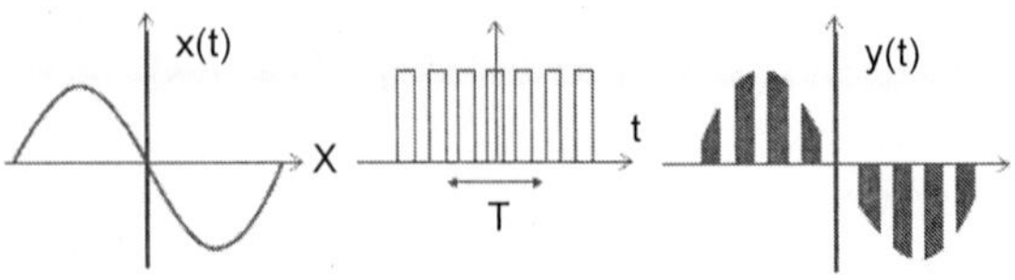

Flat Top Sampling

It is also known as practical sampling, is performed by multiplying the provided signal, $x(t)$, by an infinitely long pulse train, but the magnitude of each pulse remains fixed. A sample-and-hold procedure can be used to accomplish this sampling. The Nyquist rate determines the sampling frequency.

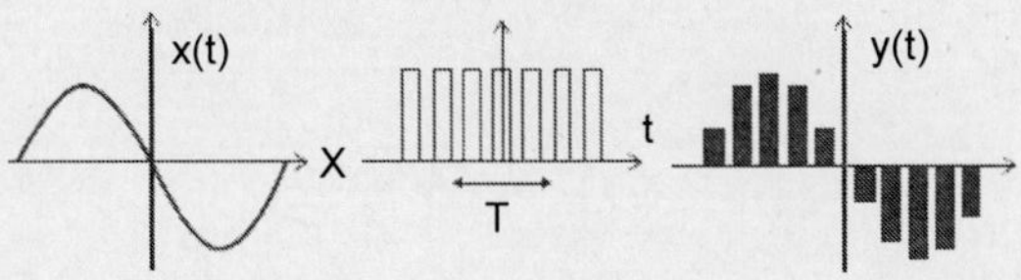

Nyquist Theorem

If and only if the sampling rate is more than or equivalent to twice of the highest frequency component of the given signal, an analog signal can be digitized without aliasing errors.

Pulse Amplitude Modulation

Analog data are transmitted using pulse modulation. Continuous wave shapes are sampled periodically in this system. Only at sampling periods, together with synchronization signals, is signal information delivered. The original waves can be reconstructed from the sample data received at the reception point. PAM is one of the most straightforward modulation techniques. It is a modulation technique that samples the input signal periodically and makes each sample proportionate to the magnitude of the signal at the moment it is sampled. The pulses are subsequently transferred via wires or cables to the modulated carrier.

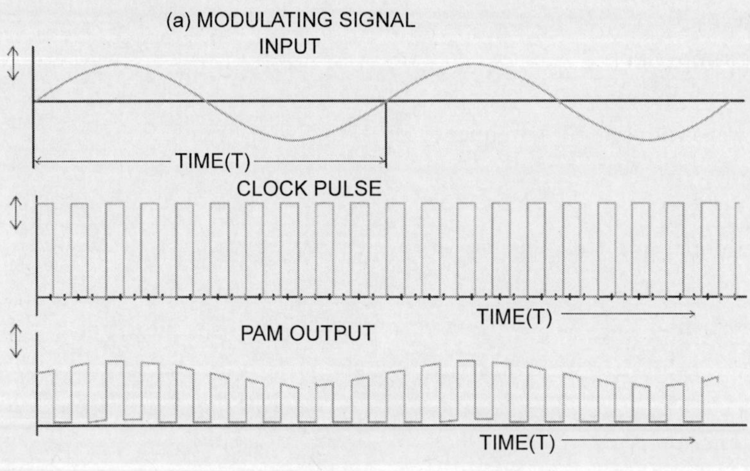

CODE

Implementation and Verification of the Nyquist Sampling Theorem

```
clc;
clear all;
time_int=-20:.01:20;
Time_p=5;
msg_freq=1/Time_p;
msg_sig=cos(2*pi*msg_freq*time_int);
subplot(2,2,1);
plot(time_int,msg_sig);
xlabel('time');ylabel('x(t)');
```

```
title('Input wave');
grid;
discret_t1=-5:1:5;
samp_freq1=1.6*msg_freq;
samp_freq2=2*msg_freq;
samp_freq3=8*msg_freq;
samp_sig1=cos(2*pi*msg_freq/samp_freq1*discret_t1);
subplot(2,2,2);
stem(discret_t1,samp_sig1);
xlabel('t');ylabel('samples');
title('Discrete wave with sampling frequency less than Nyquist rate');
hold on
subplot(2,2,2);
plot(discret_t1,samp_sig1)
dicret_t2=-6:1:6;
samp_sig2=cos(2*pi*msg_freq/samp_freq2*dicret_t2);
subplot(2,2,3);
stem(dicret_t2,samp_sig2);
xlabel('t');ylabel('samples');
title('Discrete wave with sampling frequency equal to Nyquist rate');
hold on
subplot(2,2,3);
plot(dicret_t2,samp_sig2);
grid;
discrete_t3=-40:1:40;
samp_sig3=cos(2*pi*msg_freq/samp_freq3*discrete_t3);
subplot(2,2,4);
stem(discrete_t3,samp_sig3);
xlabel('t');ylabel('samples');
title('Discrete wave with sampling frequency greater than Nyquist rate');
hold on
subplot(2,2,4);
plot(discrete_t3,samp_sig3)
```

Pulse Amplitude Modulation

```
clc;
clear all;
time_r = 0 : 1/1e3 : 20;
del = 0 : 1/5 : 20;
%message signal
```

```
msg_sig = 4+sin(4*pi/4*time_r);
figure(1);
subplot(3,1,1);
plot(time_r,msg_sig);
title(‘message’);
xlabel(‘time’);
ylabel(‘amplitude’);
%generation of pulse input
pulse_sig = pulstran(time_r,del,’rectpuls’,0.1);
subplot(3,1,2)
plot(time_r,pulse_sig);
title(‘Pulse Input ‘);
xlabel(‘time’);
ylabel(‘amplitude’);
% PAM output
PAM_sg=msg_sig.*pulse_sig;
subplot(3,1,3)
plot(time_r,PAM_sg);
title(‘PAM modulation ‘);
xlabel(‘time’);
ylabel(‘amplitude’);
```

OUTPUTS

Verification of the Sampling Theorem

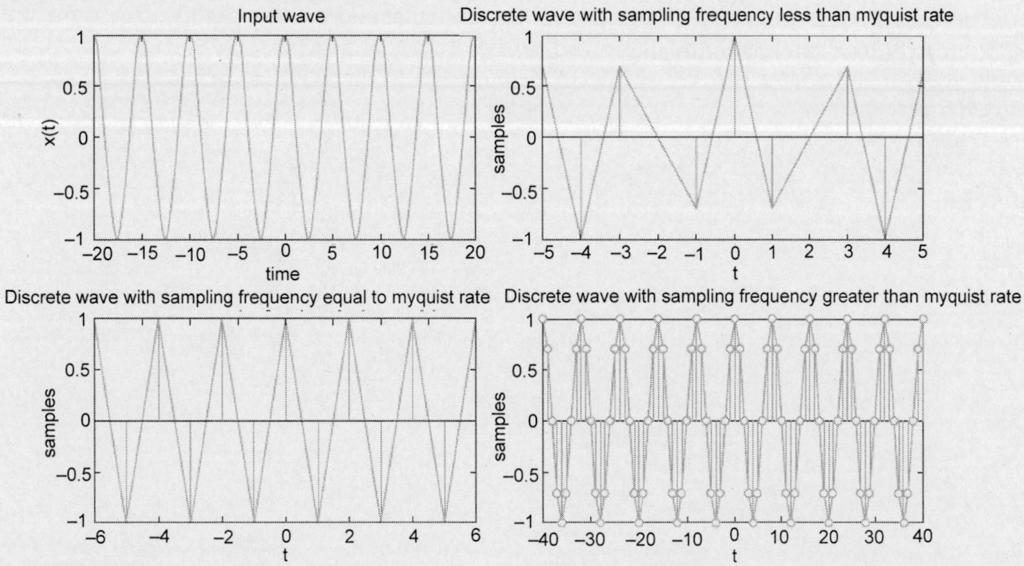

Pulse Amplitude Modulation

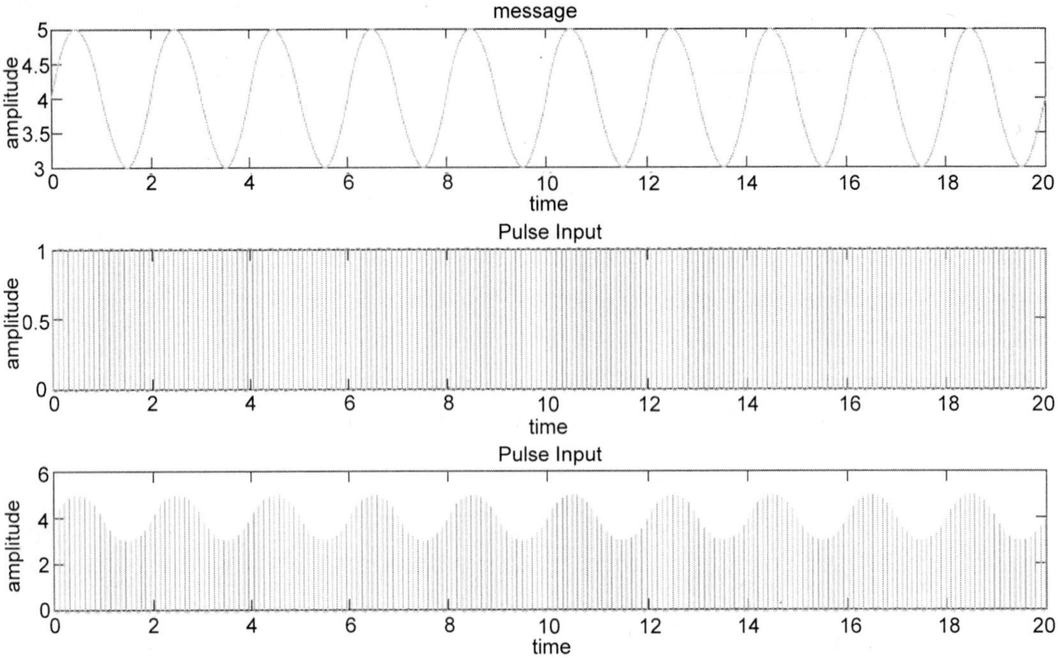

Result

Nyquist sampling theorem was verified and the PAM signal was simulated.

Pulse-Code Modulation

Objective

To perform pulse-code modulation on the given message signal and retrieve the message signal by the corresponding (Pulse-Code) demodulation.

Software Required

MATLAB 2023

Theory

Pulse Code Modulation

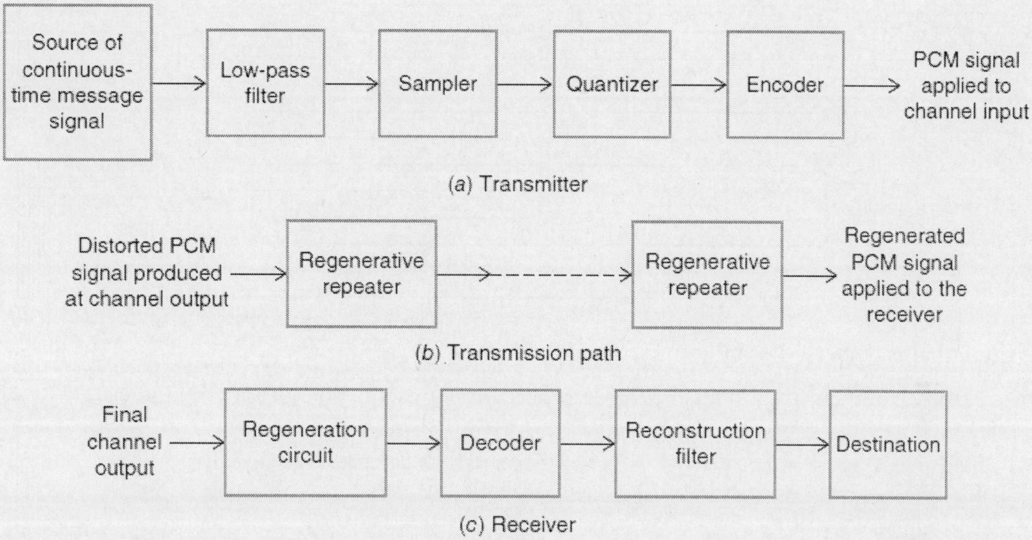

It is one of the several kinds of pulse modulations. The analog message signal is transformed into digital and sent. The digital information is demodulated and transformed back to analog at the receiver. PCM is commonly used as an analog-to-digital conversion method. The block diagram below depicts the various stages in the PCM system.

Low-Pass Filter

It is used to preserve just the frequency components that are lower than the signal's highest harmonic to avoid undesirable higher frequency components that may cause aliasing in sampling.

Sampler

Sampling is converting a continuous-time signal (analog) into a discrete-time signal with little information loss. This is done because it is tough to process all of the signal's values throughout time. Hence, sampling is performed to facilitate processing. Sampling is classified into three types - impulse sampling, natural sampling, and flat-top sampling, which are discussed in experiment 6.

Quantizer

After getting samples of the provided continuous-time signal, the magnitude is discretized. We assign quantization levels and consolidate every sample into the defined quantization quantities since the samples may possess any real value that would be impossible to transmit. Signals might be evenly spread over both the highest and lowest magnitudes or non-uniformly distributed across the magnitude range. As a result, two alternative forms of quantization are performed depending on the signal's format.

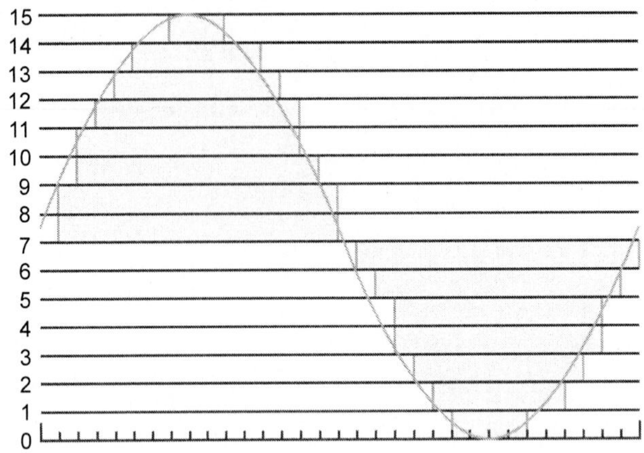

If the signal is dispersed equally,

- Number of quantization levels, $L = 2^n$, where 'n' represents the number of bits.
- Step size, $SS = \frac{\max(x(t)) - \min(x(t))}{L-1}$, where '$x(t)$' is the message signal.
- Maximum quantization error, $\Delta = \frac{SS}{2}$, where SS is the step size.
- *Quantization levels* = $\{\min(x(t)), \min(x(t)) + SS, \ldots, \min(x(t)) + (L-1).SS\}$
- *Quantization values* = $\{\min(x(t)) + \Delta, \min(x(t)) + \Delta + SS, \ldots, \min(x(t)) + \Delta + (L-2).SS\}$

The digitized value in between two successive levels is assigned $x(t)$ values across the levels. The error caused by allocating digitized values to data is known as quantization error. When the signal is not equally distributed, a more comprehensive range of levels is provided where most of the data is concentrated.

Encoder

Because quantization stages take real values, they are assigned for digital transmission. Binary coding is one of the most common types of coding. Based on the total number of quantization stages, $L = 2^n$, each stage is given a binary code ranging from 0 to $2^n - 1$ in binary format. These 0s and 1s are sent across the communication channel as pulses.

Regenerative Repeaters

Numerous regenerative replicators exist in the signal's propagation path because noise accumulates to the signal upon transmission. The circuit for a regenerative repeater is given in the graphic below. The equalizer is used to form warped pulses with phase and magnitude abnormalities. A clock circuit generates the pulses depending on the equalizer output. The decision-making device decides if the sum of the magnitudes of the digitized pulse and noise exceeds a predefined voltage level.

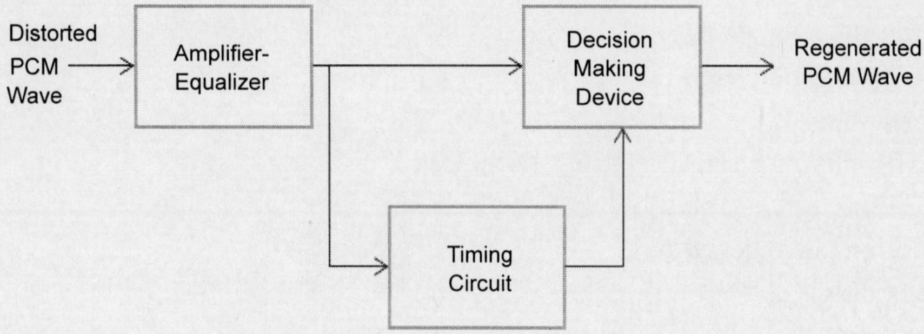

Decoder

The decoder converts values into quantization levels according to the bits received. The decoder must be supplied with the number of bits utilized in encoding.

Reconstruction Filter

During digital-to-analog conversion, a reconstruction filter creates a uniform analog output from a digital input.

CODE

```
clc
clear all;
close all;
samp_freq=4000;
sig_freq = 100;
msg_amp=2;
num_bits= 3;
%Number of quantisation levels
num_levels=2^num_bits;
%Time period of the message signal
time_p=1/sig_freq;
%Range of time axis
time_r=0:1/samp_freq:10/sig_freq;
%Length of time axis
time_r_n=length(time_r);
%Range of frequency axis
freq_r=0:samp_freq/time_r_n:((time_r_n-1)/time_r_n*samp_freq);
%message signal
msg_sig=msg_amp*sin(2*pi*sig_freq*time_r);
%Defining Step size
step_size=(max(msg_sig)-min(msg_sig))/(num_levels-1);
```

```
del=step_size/2;
quant_lev=zeros(1,num_levels);
%represents quantisation levels
for i=1:num_levels
  quant_lev(i)=min(msg_sig)+((i-1)*step_size);
end
quant_val=zeros(1,num_levels-1);
%Quantised values
for i=1:num_levels-1
  quant_val(i)=min(msg_sig)+del+(i-1)*step_size;
end
%quantising each value in 'm' to quantised values
msg_quant_val=zeros(1,time_r_n);
%encoded signal
encoded_sig=zeros(1,num_bits*time_r_n);
t1=linspace(0,time_p,length(encoded_sig));
for i=1:time_r_n
  k=((i-1)*num_bits)+1;
  if msg_sig(i)>quant_val(num_levels-1)
    msg_quant_val(i)=quant_lev(num_levels);
    s=de2bi(num_levels-1,num_bits);
    for d=1:num_bits
      encoded_sig(k+d-1)=s(d);
    end
  elseif msg_sig(i)<=quant_val(1)
    msg_quant_val(i)=quant_lev(1);
    s=de2bi(0,num_bits);
    for d=1:num_bits
      encoded_sig(k+d-1)=s(d);
    end
  else
    for j=1:num_levels-2
      if((msg_sig(i)>quant_val(j))&&(msg_sig(i)<=quant_val(j+1)))
        msg_quant_val(i)=quant_lev(j+1);
        s=de2bi(j,num_bits);
        for d=1:num_bits
          encoded_sig(k+d-1)=s(d);
```

```
        end
      end
    end
  end
end
%quantisation noise
quant_noise=msg_sig-msg_quant_val;
disp('maximum of Quantixation Noise');
disp(max(quant_noise));
disp('del');
disp(del);
Signal_Power=var(msg_sig);
noise_var=(step_size.^2)/12;
SQNR_th=Signal_Power/noise_var;
SQNR_dB_th=10*log(SQNR_th);
disp('Theoretical SQNR');
disp(SQNR_th);
disp('Theoretical SQNR in dB');
disp(SQNR_dB_th);
Quantisation_noise_var=var(quant_noise);
SQNR_p=Signal_Power/Quantisation_noise_var;
SQNR_dB_p=10*log(SQNR_p);
disp('Simulated SQNR');
disp(SQNR_p);
disp('SImulated SQNR in dB');
disp(SQNR_dB_p);

figure(1)
plot(time_r,msg_sig);
title('message and quantized signal');
axis([0 time_p min(msg_sig)-1 max(msg_sig)+1]);
grid on;
hold on;
plot(time_r,msg_quant_val,'r');
hold off;
figure(2);
subplot(2,1,1);
```

```
plot(time_r,msg_sig);
grid on;
hold on;
plot(time_r,msg_quant_val,'r');
title('message and digitized signal');
hold off;
n1=length(encoded_sig);
subplot(2,1,2);
stem(encoded_sig);
axis([0 n1 -1 2]);
title('encoded signal');
grid on;
figure(3);
plot(time_r,quant_noise);
title('Quantised noise signal');
%cut off frequency of the filter
fc=sig_freq+10;
% order of the filter
N=201;
%cut off frequency as limits for the filter
w=2*pi*fc/samp_freq;
%range of filter
n=-100:1:100;
%Filter characteristic
h=2*fc/samp_freq*sinc(2*fc*n/samp_freq);
%filtered signal
y=filter(h,1,msg_quant_val);
figure(4);
plot(n,h);
title('low pass filter characteristic function');
figure(5)
subplot(3,2,1);
plot(time_r,msg_sig);
title('message signal');
subplot(3,2,2);
plot(freq_r,(fft(msg_sig)));
title('Spectrum of message signal');
```

```
subplot(3,2,3);
plot(time_r,msg_quant_val);
title('quantised message signal');
subplot(3,2,4)
plot(freq_r,(fft(msg_quant_val)));
title('Spectrum of digitised message signal');
subplot(3,2,5);
plot(time_r,y);
title('Filtered signal');
subplot(3,2,6);
plot(freq_r,abs(fft(y)));
title('Spectrum of filtered signal');
```

OUTPUTS

Command window

Maximum of quantization noise	0.2857
Delta (Δ)	0.2857
Theoretical SQNR	73.5000
Theoretical SQNR in dB	42.9729
Simulated SQNR	76.5918
Simulated SQNR in dB	43.3849

Plots

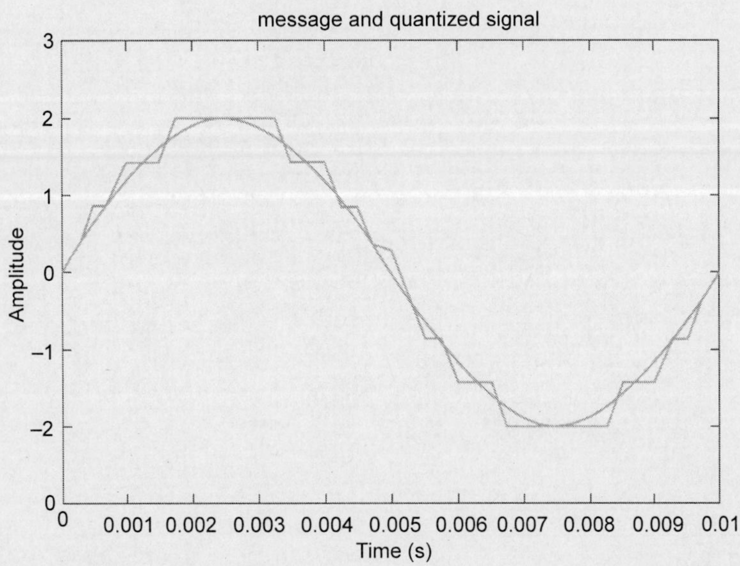

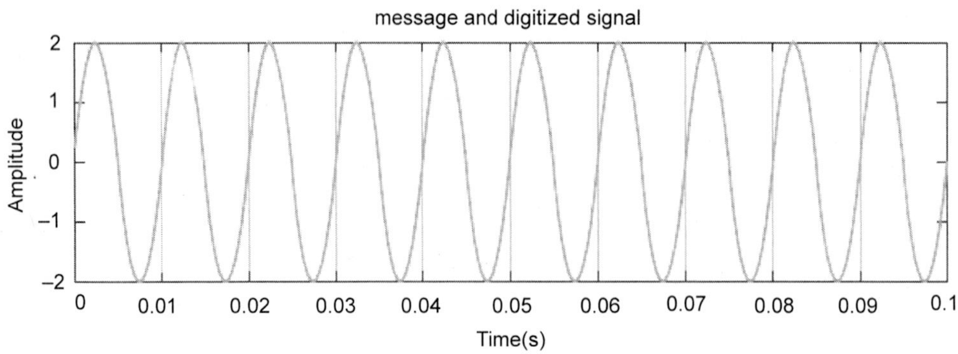
message and digitized signal
Amplitude
Time(s)

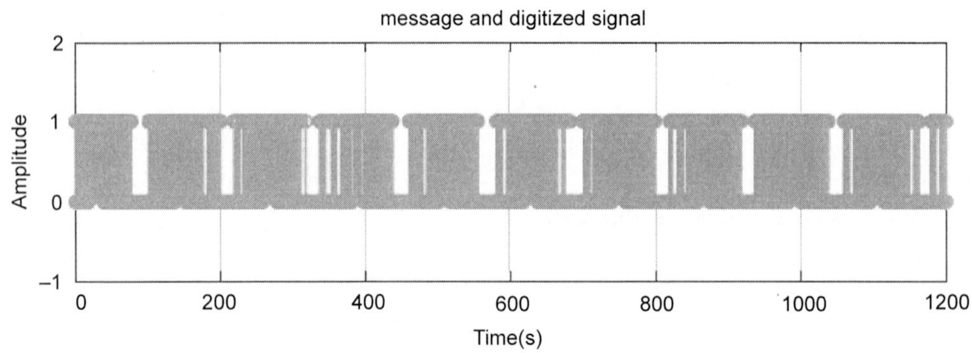
message and digitized signal
Amplitude
Time(s)

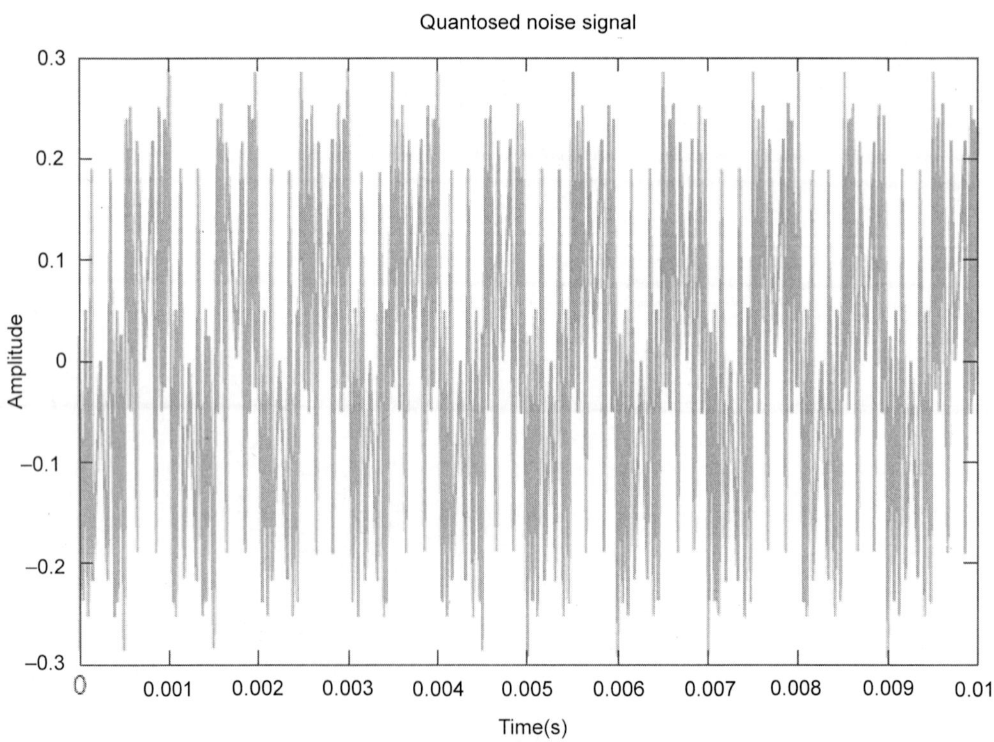
Quantosed noise signal
Amplitude
Time(s)

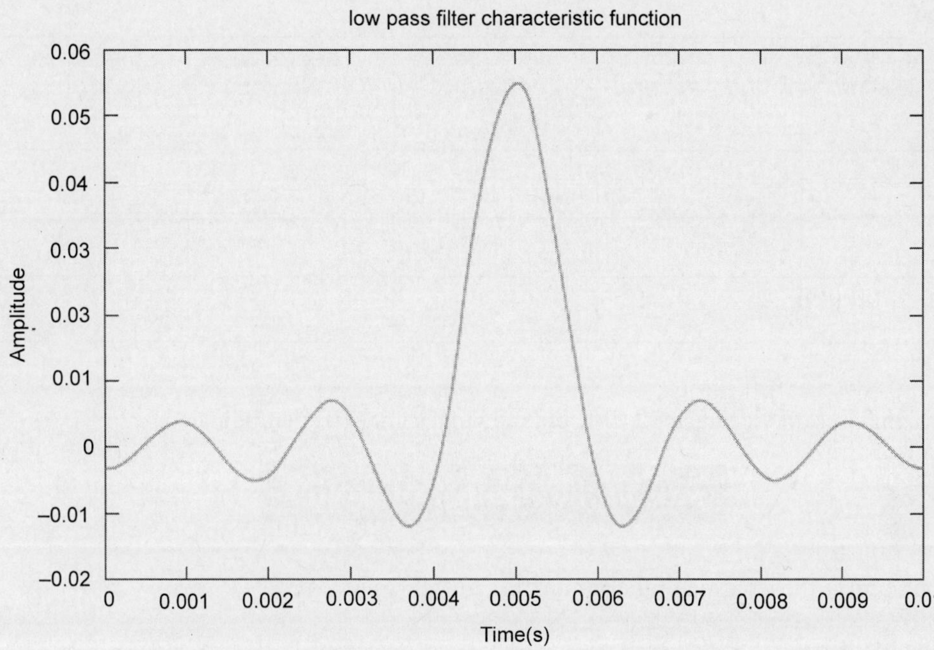
low pass filter characteristic function
Amplitude
0.06
0.05
0.04
0.03
0.03
0.01
0
−0.01
−0.02
0
0.001
0.002
0.003
0.004
0.005
0.006
0.007
0.008
0.009
0.01
Time(s)

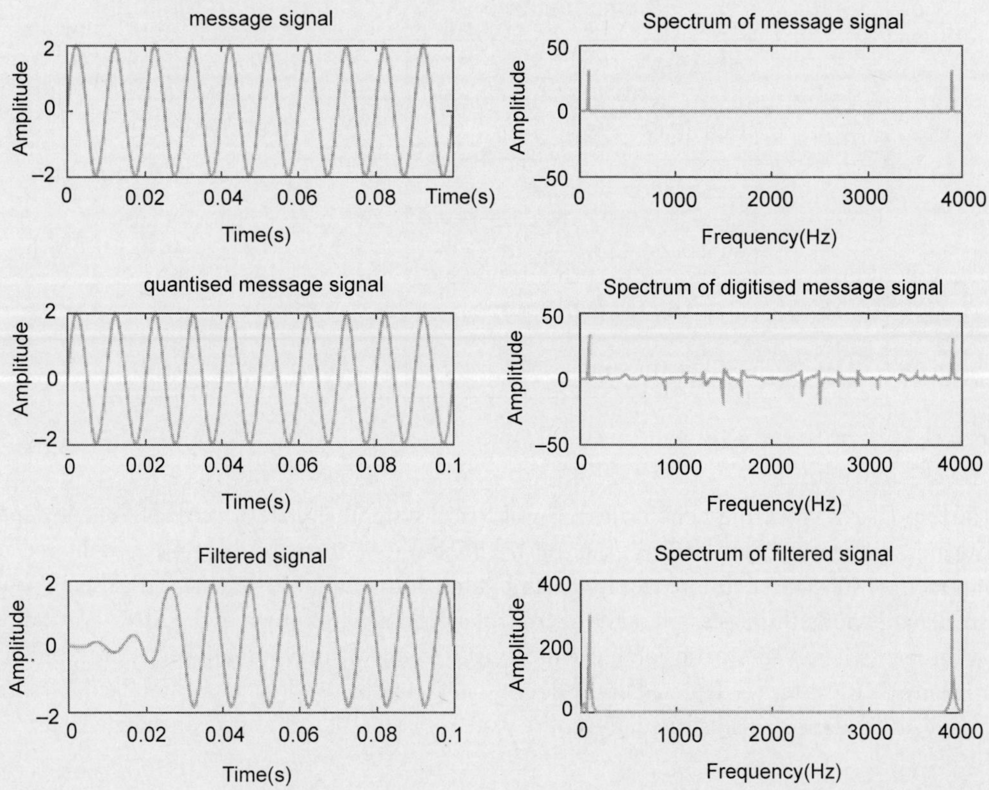
message signal
Amplitude
Time(s)
Spectrum of message signal
Frequency(Hz)
quantised message signal
Time(s)
Spectrum of digitised message signal
Frequency(Hz)
Filtered signal
Time(s)
Spectrum of filtered signal
Frequency(Hz)

Observation

No. of quantization levels	Maximum quantization error	del = step size/2	Theoretical SQNR (dB)	Simulated SQNR (dB)

Results

The analog signal was digitized using PCM, and the same signal was demodulated.

Automatic Gain Controller

Objective

To study and observe the working of the automatic gain controller (AGC).

Components Required

S. No.	Components	Range/Part No.
1.	OP-AMP	OPA547
2.	Resistor	10 kΩ (2)
3.	Capacitors	10 μF
4.	Multipliers (**X**)	2
5.	Cathode ray oscilloscope	–
6.	DC power supply	–

Theory

Automatic Gain Controller

The response of a measurement device in an electrical system's signal path could fluctuate depending on the degree of intensity of the feed. We can construct the amplifier so that its gain can be altered actively to accommodate substantial differences in the amplitude of the input. When the signal being processed has a small bandwidth, this is feasible, and the control mechanism is known as the AGC. We additionally employ the abbreviation automatic volume control (AVC) since we may want to keep the amplifier's final output steady. AGC system is depicted in the figure. The figure below depicts the standard I/O attributes of an AGC/AVC system. As illustrated, the entire system's resultant value stays unchanged.

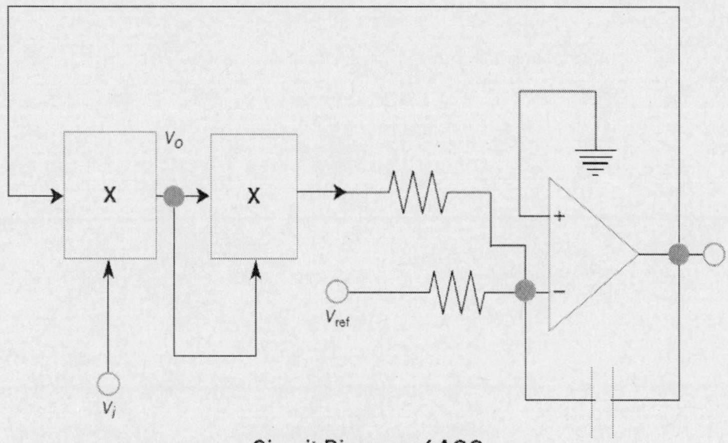

Circuit Diagram of AGC

Circuit diagram of AGC

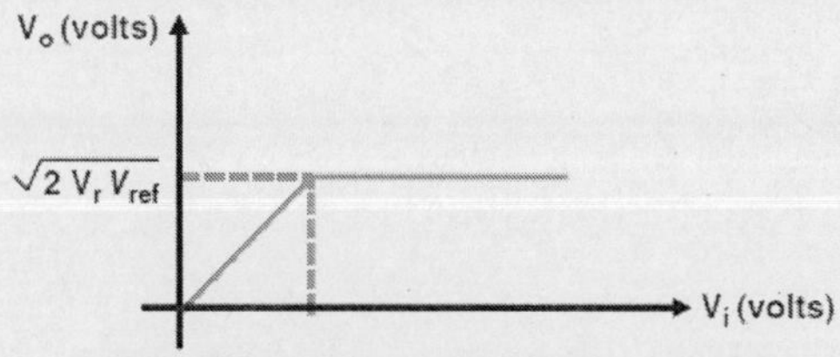

Input-output characteristics of AGC

Input–output characteristics of AGC

Procedure

- Build the AGC circuit as shown in the schematic.
- Give input from the function generator and take the output from the oscilloscope.
- Plot the output as a function of the input potential.

Observation

S. No.	INPUT VOLTAGE (V_i)	OUTPUT VOLTAGE (V_o)	CONTROL VOLTAGE (V_{ref})
1.			
2.			
3.			

Result

The automatic gain controller was constructed and studied.

CO and PO Attainment Table

Course outcomes (COs) for this course can be mapped with the programme outcomes (POs) after the completion of the course and a correlation can be made for the attainment of POs to analyze the gap. After proper analysis of the gap in the attainment of POs necessary measures can be taken to overcome the gaps.

Table for CO and PO Attainment

Course Outcomes	Attainment of Programme Outcomes (1- Weak Correlation; 2- Medium correlation; 3- Strong Correlation)						
	PO-1	PO-2	PO-3	PO-4	PO-5	PO-6	PO-7
CO-1							
CO-2							
CO-3							
CO-4							
CO-5							
CO-6							

The data filled in the above table can be used for gap analysis.

INDEX

❑❑